W9-BGQ-774

# ARRL's
# GENERAL
# Q&A

## Upgrade to a General Class Ham Radio License!

NEW!
**Fourth Edition**

**By Ward Silver, NØAX**

**Contributing Editor:**
Mark Wilson, K1RO

**Editorial Assistant:**
Maty Weinberg, KB1EIB

**Production Staff:**
David Pingree, N1NAS, Senior Technical Illustrator
Jodi Morin, KA1JPA, Assistant Production Supervisor: Layout
Sue Fagan, KB1OKW, Graphic Design Supervisor: Cover Design
Michelle Bloom, WB1ENT, Production Supervisor

**Front Cover Photo: G. Hilton Dean, W4GHD**

**ARRL** *The national association for* **AMATEUR RADIO**®
225 Main Street, Newington, CT 06111-1494

**www.arrl.org**

This book may be used for General license exams given beginning July 1, 2011. *QST* and the ARRL website (**www. arrl.org**) will have news about any rules changes affecting the General class license or any of the material in this book.

We strive to produce books without errors. Sometimes mistakes do occur, however. When we become aware of problems in our books (other than obvious typographical errors), we post corrections on the ARRL website. If you think you have found an error, please check **www.arrl.org** for corrections. If you don't find a correction there, please let us know either by using the Feedback Form at the back of this book or by sending e-mail to **pubsfdbk@arrl.org**.

# Contents

# Foreword

*Congratulations and welcome!* If you have decided to upgrade to the General class Amateur Radio license or are thinking about giving it a try, this book will help. Upgrading to General will open up the world of shortwave communications and a wide range of Amateur Radio bands and modes and activities — you won't regret it! For more than 90 years, the ARRL has helped Amateur Radio operators get the most out of their hobby by advancing their operating and technical skills. You'll find the ARRL license preparation materials to be the most complete package available — designed to help you do much more than just pass the written exam.

All of the information you need to pass the Element 3 General class written exam is here in the fourth edition of *ARRL's General Q&A*. Each and every question in the General class examination question pool is addressed with study material that explains the correct answer and provides background information. If you can answer the questions in this book, you can pass the written exam with confidence.

In addition to this book, *The ARRL General Class License Manual* is a detailed text reference that helps you understand the electronics theory, operating practices and FCC rules. This not only helps you pass the exam, but makes you more confident on the air and in your "shack." The manual includes a CD-ROM with software to help you practice for the exam — it contains all of the questions with a short explanation and helpful graphics.

If you'd like to take part in one of Amateur Radio's great traditions (and have a lot of fun, too), the ARRL provides complete training materials for learning the Morse code: audio CDs, training software, Morse keys, and even a code-practice oscillator kit. While Morse proficiency is no longer required for you to get an Amateur Radio license, it remains quite popular with amateurs for its efficiency and simple elegance.

Visit the ARRL's comprehensive website at **www.arrl.org** and browse the latest news about Amateur Radio, tap into the wealth of services provided by the ARRL and look through the ARRL online catalog for publications and supplies to support any activity.

The fourth edition of *ARRL's General Q&A* is the product of cooperation between readers of the ARRL study materials and the ARRL staff. You can help make this book better by providing your own feedback. After you have passed your exam, write your suggestions, questions and comments on the Feedback Form at the back of the book then send the form to us. Comments from readers are very important in making subsequent editions more effective and useful to readers.

Thanks for making the decision to upgrade — we hope to hear you on the air soon, using your new General class privileges and enjoying more of Amateur Radio. Good luck!

David Sumner, K1ZZ
Chief Executive Officer
Newington, Connecticut
March 2011

New Ham Desk
ARRL Headquarters
225 Main Street
Newington, CT
06111-1494
860-594-0200

Prospective new amateurs call:
**800-32-NEW-HAM (800-326-3942)**

You can also contact us via e-mail:
**newham@arrl.org**

or check out
**www.arrl.org**

# What is
# Amateur Radio?

Perhaps you've just picked up this book in the library or from a bookstore shelf and are wondering what this Amateur Radio business is all about. Maybe you have a friend or relative who is a "ham" and you're interested in becoming one, as well. In that case, a short explanation is in order.

Amateur Radio or "ham radio" is one of the longest-lived wireless activities. Amateur experimenters were operating right along with Marconi in the early part of the 20th century. They have helped advance the state-of-the-art in radio, television, wireless data and dozens of other communications services right up to the present day. There are nearly 700,000 amateur radio operators or "hams" in the United States alone and several million more around the world!

Amateur Radio in the United States is a formal *communications service*, administered by the Federal Communications Commission or FCC. Created officially in its present form in 1934, the amateur service is intended to foster electronics and radio experimentation, provide emergency backup communications, encourage private citizens to train and practice operating, and even spread the goodwill of person-to-person contact over the airwaves.

## Who Is a Ham and What Do Hams Do?

Anyone can be a ham — there are no age limits or physical requirements that prevent anyone from passing their license exam and getting on the air. Kids as young as 6 years old have passed the basic exam and there are hams over the age of 100. You probably fall somewhere in the middle of that range.

Once you get on the air and start meeting other hams, you'll find a wide range of capabilities and interests. Of course, there are many technically skilled hams who work as engineers, scientists or technicians. But just as many don't have a deep technical background. You're just as likely to encounter writers, public safety personnel, students, farmers, truck drivers — anyone with an interest in personal communications over the radio.

The activities of Amateur Radio are incredibly varied. Amateurs who hold the Technician class license — the first license for hams in the US — communicate primarily with local and regional amateurs using relay stations called *repeaters*. Known as "Techs," they sharpen their skills of operating while portable and mobile, often joining emergency communications teams. They may instead focus on the burgeoning wireless data networks assembled and used by hams around the world. Techs can make use of the growing number of Amateur Radio satellites, built and launched by hams along with the commercial "birds." Technicians transmit their own television signals, push the

**Field Day is the largest event in Amateur Radio as thousands of North American hams practice teamwork and responding to communications emergencies by operating from portable stations.**

limits of signal propagation through the atmosphere and experiment with microwaves. Hams hold most of the world records for long-distance communication on microwave frequencies, in fact!

Hams who advance or *upgrade* to General class are granted additional privileges to use the frequencies usually associated with shortwave operation. This is the traditional Amateur Radio you may have encountered in movies or books. On these frequencies, signals travel worldwide and so General and Extra class amateurs can make direct contact with foreign hams. No Internet, phone systems, or data networks are required. It's just you, your radio, and the ionosphere — the upper layers of the Earth's atmosphere!

**Amateur Radio emergency communications or "emcomm" teams worked together with state Health Department personnel in this 2010 emergency response training exercise.**

Many hams use voice, Morse code, and computer data signals to communicate, engaging in conversation, participating in competitions, and exchanging text and images. All of these signals are mixed together on the frequencies where hams operate, making the experience of tuning a radio receiver through the crowded bands an interesting experience.

One thing common to all hams is that all of their operation is noncommercial, especially the volunteers who provide emergency communications. Hams pursue their hobby purely for personal enjoyment and to advance their skills, taking satisfaction from providing services to their fellow citizens. This is especially valuable after natural disasters such as hurricanes and earthquakes when commercial systems are knocked out for a while. Amateur operators rush in to provide backup communication for hours, days, weeks or even months until the regular systems are restored. All this from a little study and a simple exam!

## Want to Find Out More?

If you'd like to find out more about Amateur Radio in general, there is lots of information available on the Internet. A good place to start is on the American Radio Relay League's (ARRL) ham radio introduction page, **www.arrl.org/new-to-ham-radio**. Books like *Ham Radio for Dummies* and *Getting Started With Ham Radio* will help you "fill in the blanks" as you learn more.

Supported by books and web pages, there is no better way to learn about ham radio than to meet your local amateur operators. It is quite likely that no matter where you live in the United States, there is a ham radio club in your area — perhaps several! The ARRL provides a club lookup Web page at **www.arrl.org/find-a-club** where you can find a club just by entering your Zip code or state. Carrying on the tradition of mutual assistance, many clubs make helping newcomers to ham radio a part of their charter.

If it sounds like hams are confident that you'll find their activities interesting, you're right! Amateur Radio is much more than just talking on a radio as you'll find out. It's an opportunity to dive into the fascinating world of radio communications, electronics and computers as deeply as you wish to go. Welcome!

# When to Expect
## New Books

A Question Pool Committee (QPC) consisting of representatives from the various Volunteer Examiner Coordinators (VECs) prepares the license question pools. The QPC establishes a schedule for revising and implementing new Question Pools. The current Question Pool revision schedule is as follows:

| Question Pool | Current Study Guides | Valid Through |
|---|---|---|
| **Technician**<br>(Element 2) | *The ARRL Ham Radio<br>License Manual*, 2nd Edition<br>*ARRL's Tech Q&A*, 5th Edition | June 30, 2014 |
| **General**<br>(Element 3) | *The ARRL General Class<br>License Manual*, 7th edition<br>*ARRL's General Q&A*, 4th Edition | June 30, 2015 |
| **Amateur Extra**<br>(Element 4) | *The ARRL Extra Class<br>License Manual*, 9th Edition<br>*ARRL's Extra Q&A*, 2nd Edition | June 30, 2012 |

As new question pools are released, ARRL will produce new study materials before the effective date of the new pools. Until then, the current question pools will remain in use and current ARRL study materials, including this book, will help you prepare for your exam.

As the new question pool schedules are confirmed, the information will be published in *QST* and on the ARRL website at **www.arrl.org**.

# How to Use
## This Book

To earn a General class Amateur Radio license, you must pass (or receive credit for) FCC Elements 2 (Technician class) and 3 (General class). This book is designed to help you prepare for and pass the Element 3 written exam. If you do not already have a Technician class license, you will need study materials for the Element 2 (Technician) exam.

The Element 3 exam consists of 35 questions about Amateur Radio rules, theory and practice, as well as some basic electronics. A passing grade is 74%, so you must answer 26 of the 35 questions correctly.

*ARRL's General Q&A* has 10 sections that follow subelements G1 through G0 in the General class syllabus. The questions and multiple choice answers in this book are printed exactly as they were written by the Volunteer Examiner Coordinators' Question Pool Committee, and exactly as they will appear on your exam. (Be careful, though. The position of the answers may be scrambled from that of the actual exam so you won't be able to simply memorize an answer letter for each question.) In this book, the letter of the correct answer is printed in **boldface** type just before the explanation.

The ARRL also maintains a special web page for General class students at **www.arrl.org/general-class-license-manual**. The useful and interesting on-line references and other supplemental information listed there put you one click away from related and useful information. Be sure to visit that web page during your preparation for the exam.

If you are taking a licensing class, help your instructors by letting them know about areas in which you need help. They want you to learn as thoroughly and quickly as possible, so don't hold back with your questions. Similarly, if you find the material particularly clear or helpful, tell them that, too, so it can be used in the next class!

## What We Assume About You

You don't have to be a technical guru or an expert operator to upgrade to General class! As you progress through the material, you'll build on the basic science of radio and electricity you mastered for Technician. No advanced mathematics is introduced and if math gives you trouble, tutorials are listed at **www.arrl.org/general-class-license-manual**. As with the Technician license, mastering rules and regulations will require learning some new words and a few numbers. You should have a basic calculator, which you'll also be allowed to use during the license exam.

If you have some background in radio, perhaps as a technician or trained operator, you may be able to skip over some of the sections. It's common for technically-minded students to focus on the rules and regulations while students with an operating background tend to need the technical material more. Whichever you may be, be sure that you can answer the questions because any of them may be on the test!

ARRL's *General Q&A* can be used either by an individual student, studying on his or her own, or as part of a licensing class taught by an instructor. If you're part of a class, the instructor will set the order in which the material is covered. The solo student can move at any pace and in any convenient order. You'll find that having a buddy to study with makes learning the material more fun as you help each other over the rough spots.

Don't hesitate to ask for help! If you can't find the answer in the book or at the website, email your question to the ARRL's New Ham Desk, **newham@arrl.org**. The ARRL's experts will answer directly or connect you with another ham that can answer your questions.

## Online Exams

While you're studying and when you feel like you're ready for the actual exam you can get some good practice by taking one of the on-line Amateur Radio exams. These websites use the same question pool to construct an exam with the same number and variety of questions that you'll encounter on exam day. The exams are free and you can take them over and over again in complete privacy. Links to on-line exams can be found at **www.arrl.org/general-class-license-manual**. The *ARRL General Class License Manual* also includes a CD-ROM with exam practice software.

These exams are quite realistic and you get quick feedback about the questions you didn't answer correctly. When you find yourself passing the practice exams by a comfortable margin, you'll be ready for the real thing! A note of caution, be sure that the practice exam uses questions from the *current* question pool.

# Testing
## Process

When you're ready, you'll need to find a test session. If you're in a licensing class, the instructor will help you find and register for a session. Otherwise, you can find a test session by using the ARRL's web page for finding exams, **www.arrl.org/find-an-amateur-radio-license-exam-session**. If you can register for the test session in advance, do so. Other sessions, such as those at hamfests or conventions, are available to anyone that shows up or walk-ins. You may have to wait for an available space though, so go early!

As for all amateur exams, the General class exam is administered by Volunteer Examiners (VEs). All VEs are certified by a Volunteer Examiner Coordinator (VEC) such as the ARRL/VEC. This organization trains and certifies VEs and processes the FCC paperwork for their test sessions.

Bring your current license original and a photocopy (to send with the application). You'll need two forms of identification including at least one photo ID, such as a driver's license, passport or employer's identity card. Know your Social Security Number (SSN). You can bring pencils or pens, blank scratch paper and a calculator, but any kind of computer or on-line device is prohibited.

The FCC allows Volunteer Examiners (VEs) to use a range of procedures to accommodate applicants with various disabilities. If this applies to you, you'll still have to pass the test, but special exam procedures can be applied. Contact your local VE team or the Volunteer Examiner Coordinator (VEC) responsible for the test session you'll be attending. Or contact the ARRL/VEC Office at 225 Main St, Newington CT 06111-1494 or by phone at 860-594-0200.

Once you're signed in, you'll need to fill out a copy of the National Conference of Volunteer Examiner Coordinator's (NCVEC) Quick Form 605. This is an application for a new or upgraded license. It is used only at test sessions and for a VEC to process a license renewal or a license change. Do not use an NCVEC Quick Form 605 for any kind of application directly to the FCC — it will be rejected. Use a regular FCC Form 605. After filling out the form, pay the current test fee and get ready.

# NCVEC QUICK-FORM 605 APPLICATION FOR AMATEUR OPERATOR/PRIMARY STATION LICENSE

## SECTION 1 - TO BE COMPLETED BY APPLICANT

| PRINT LAST NAME | SUFFIX (Jr., Sr.) | FIRST NAME | INITIAL | STATION CALL SIGN (IF ANY) |
|---|---|---|---|---|
| Grimaldi | | Amanda | E | KB1KJC |

| MAILING ADDRESS (Number and Street or P.O. Box) | SOCIAL SECURITY NUMBER (SSN) or (FRN) FCC FEDERAL REGISTRATION NUMBER |
|---|---|
| 225 Main St. | 0005189337 |

| CITY | STATE CODE | ZIP CODE (5 or 9 Numbers) | E-MAIL ADDRESS (OPTIONAL) |
|---|---|---|---|
| Newington | CT | 06111 | |

| DAYTIME TELEPHONE NUMBER (Include Area Code) OPTIONAL | FAX NUMBER (Include Area Code) OPTIONAL | ENTITY NAME (IF CLUB, MILITARY RECREATION, RACES) |
|---|---|---|
| 860-594-0200 | | |

Type of Applicant: ☒ Individual ☐ Amateur Club ☐ Military Recreation ☐ RACES (Modify Only)

| CLUB, MILITARY RECREATION, OR RACES CALL SIGN |
|---|
| |

**I HEREBY APPLY FOR** (Make an X in the appropriate box(es))

| SIGNATURE OF RESPONSIBLE CLUB OFFICIAL (not trustee) |
|---|
| |

☐ **EXAMINATION** for a **new** license grant

☒ **EXAMINATION** for **upgrade** of my license class

☐ **CHANGE** my **name** on my license to my new name

Former Name: _____
(Last name) (Suffix) (First name) (MI)

☐ **CHANGE** my mailing address to **above** address

☐ **CHANGE** my station **call sign** systematically

Applicant's Initials: _____

☐ **RENEWAL** of my license grant.

| Do you have another license application on file with the FCC which has not been acted upon? | PURPOSE OF OTHER APPLICATION | PENDING FILE NUMBER (FOR VEC USE ONLY) |
|---|---|---|
| | | |

**I certify that:**
* I waive any claim to the use of any particular frequency regardless of prior use by license or otherwise;
* All statements and attachments are true, complete and correct to the best of my knowledge and belief and are made in good faith;
* I am not a representative of a foreign government;
* I am not subject to a denial of Federal benefits pursuant to Section 5301 of the Anti-Drug Abuse Act of 1988, 21 U.S.C. § 862;
* The construction of my station will NOT be an action which is likely to have a significant environmental effect (See 47 CFR Sections 1.1301-1.1319 and Section 97.13(a));
* I have read and WILL COMPLY with Section 97.13(c) of the Commission's Rules regarding RADIOFREQUENCY (RF) RADIATION SAFETY and the amateur service section of OST/OET Bulletin Number 65.

**Signature of applicant** (Do not print, type, or stamp. Must match applicant's name above.) (Clubs: 2 different individuals must sign)

X _(signature)_ Date Signed: 01/23/2011

## SECTION 2 - TO BE COMPLETED BY ALL ADMINISTERING VEs

Applicant is qualified for operator license class:

☐ **NO NEW LICENSE OR UPGRADE WAS EARNED**

☐ **TECHNICIAN** Element 2

☒ **GENERAL** Elements 2 and 3

☐ **AMATEUR EXTRA** Elements 2, 3 and 4

| DATE OF EXAMINATION SESSION |
|---|
| 01-23-2011 |

| EXAMINATION SESSION LOCATION |
|---|
| Newington, CT |

| VEC ORGANIZATION |
|---|
| ARRL |

| VEC RECEIPT DATE |
|---|
| |

**I CERTIFY THAT I HAVE COMPLIED WITH THE ADMINISTERING VE REQUIRMENTS IN PART 97 OF THE COMMISSION'S RULES AND WITH THE INSTRUCTIONS PROVIDED BY THE COORDINATING VEC AND THE FCC.**

| 1st VEs NAME (Print First, MI, Last, Suffix) | VEs STATION CALL SIGN | VEs SIGNATURE (Must match name) | DATE SIGNED |
|---|---|---|---|
| Kay C. Craigie | N3KN | _(signature)_ | 01-23-2011 |
| PERRY T GREEN | WY1O | _(signature)_ | 01-23-2011 |
| Penny Harts | N1NAG | _(signature)_ | 01-23-2011 |

DO NOT SEND THIS FORM TO FCC – THIS IS NOT AN FCC FORM.
IF THIS FORM IS SENT TO FCC, FCC WILL RETURN IT TO YOU WITHOUT ACTION.

NCVEC FORM 605 - February 2007
FOR VE/VEC USE ONLY - Page 1

**This sample NCVEC Quick Form 605 shows how your form will look after you have completed your upgrade to General.**

**FCC 605**
**Main Form**

**Quick-Form Application for Authorization in the Ship, Aircraft, Amateur, Restricted and Commercial Operator, and General Mobile Radio Services**

Approved by OMB
3060 - 0850
See instructions for
public burden estimate

| 1) Radio Service Code: | *H A* |
|---|---|

**Application Purpose** (Select only one) (*M D*)

| 2) **NE** – New | **RO** – Renewal Only | **WD** – Withdrawal of Application |
|---|---|---|
| **MD** – Modification | **RM** – Renewal / Modification | **DU** – Duplicate License |
| **AM** – Amendment | **CA** – Cancellation of License | **AU** – Administrative Update |

| 3) | If this request is for Developmental License or STA (Special Temporary Authorization) enter the appropriate code and attach the required exhibit as described in the instructions. Otherwise enter 'N' (Not Applicable). | (*N*) D  S  N/A |
|---|---|---|
| 4) | If this request is for an Amendment or Withdrawal of Application, enter the file number of the pending application currently on file with the FCC. | File Number |
| 5) | If this request is for a Modification, Renewal Only, Renewal / Modification, Cancellation of License, Duplicate License, or Administrative Update, enter the call sign (serial number for Commercial Operator) of the existing FCC license. If this is a request for consolidation of DO & DM Operator Licenses, enter serial number of DO. Also, if filing for a ship exemption, you must provide call sign. | Call Sign/Serial # *WY1O* |
| 6) | If this request is for a New, Amendment, Renewal Only, or Renewal Modification, enter the requested expiration date of the authorization (this item is optional). | MM        DD |
| 7) | Does this filing request a Waiver of the Commission's Rules? If 'Y', attach the required showing as described in the instructions. | (*N*) Yes  No |
| 8) | Are attachments (other than associated schedules) being filed with this application? | (*N*) Yes  No |

**Applicant/Licensee Information**

| 9) FCC Registration Number (FRN): | *0005189337* |
|---|---|

10) Applicant/Licensee legal entity type: (Select One )

☑Individual   ☐ Corporation   ☐ Unincorporated Association   ☐ Trust   ☐ Government Entity
☐Consortium   ☐ General Partnership   ☐ Limited Liability Company   ☐ Limited Liability Partnership
☐Limited Partnership   ☐ Other (Description of Legal Entity) _____

| 11) First Name (if individual): *PERRY* | MI: *T* | Last Name: *GREEN* | Suffix: |
|---|---|---|---|

12) Entity Name (if other than individual):

13) If the Licensee name is being updated, is the update a result from the sale (or transfer of control) of the license(s) to another party and for which proper Commission approval has not been received or proper notification not provided?   (   ) Yes  No

14) Attention To:

| 15) P.O. Box: *53* | And/Or | 16) Street Address: |
|---|---|---|

| 17) City: *NEW HARTFORD* | 18) State: *CT* | 19) Zip Code/Postal Code: *06057* | 20) Country: |
|---|---|---|---|

| 21) Telephone Number: *860 594-0200* | 22) FAX Number: |
|---|---|

23) E-Mail Address:

**Ship Applicants/Licensees Only**

24) Enter new name of vessel: _____

**Aircraft Applicants/Licensees Only**

25) Enter the new FAA Registration Number (the N-number): _____
   **NOTE:** Do not enter the leading "N".

FCC 605 – Main Form
May 2008 - Page 1

**Portions of FCC Form 605 showing the sections you would complete for a modification of your license, such as a change of address.**

# Books to Help
## You Learn

As you study the material on the licensing exam, you will have lots of other questions about the how and why of Amateur Radio. The following references, available from your local bookstore or the ARRL (**www.arrl.org/arrl-store**) will help "fill in the blanks" and give you a broader picture of the hobby:

- *Ham Radio for Dummies* by Ward Silver, NØAX. Written for new Technician and General class licensees, this book supplements the information in study guides with an informal, friendly approach to the hobby.

- *ARRL Operating Manual.* With in-depth chapters on the most popular ham radio activities, this is your guide to nets, award programs, DXing and more. It even includes a healthy set of reference tables and maps.

- *Understanding Basic Electronics* by Walter Banzhaf, WB1ANE. Students who want more technical background about electronics should take a look at this book. It covers the fundamentals of electricity and electronics that are the foundation of all radio.

- *Basic Radio* by Joel Hallas, W1ZR. Students who want more technical background about radio theory should take a look at this book. It covers the key building blocks of receivers, transmitters, antennas and propagation.

- *ARRL Handbook.* This is the grandfather of all Amateur Radio references and belongs on the shelf of hams. Almost any topic you can think of in Amateur Radio technology is represented here.

- *ARRL Antenna Book.* After the radio itself, all radio depends on antennas. This book provides information on every common type of amateur antenna, feed lines and related topics, and practical construction tips and techniques.

# Time to
# Get Started

By following these instructions and carefully studying the material in this book, soon you'll be joining the rest of the General and Extra class licensees on the HF bands! Each of us at the ARRL Headquarters and every ARRL member looks forward to the day when you join the fun. 73 (best regards) and good luck!

# 2011-2015 General Class (Element 3) Syllabus

## SUBELEMENT G1 — COMMISSION'S RULES
### [5 Exam Questions — 5 Groups]

G1A  General Class control operator frequency privileges; primary and secondary allocations

G1B  Antenna structure limitations; good engineering and good amateur practice; beacon operation; restricted operation; retransmitting radio signals

G1C  Transmitter power regulations; data emission standards

G1D  Volunteer Examiners and Volunteer Examiner Coordinators; temporary identification

G1E  Control categories; repeater regulations; harmful interference; third party rules; ITU regions

## SUBELEMENT G2 — OPERATING PROCEDURES
### [5 Exam Questions — 5 Groups]

G2A  Phone operating procedures; USB/LSB utilization conventions; procedural signals; breaking into a QSO in progress; VOX operation

G2B  Operating courtesy; band plans, emergencies, including drills and emergency communications

G2C  CW operating procedures and procedural signals, Q signals and common abbreviations; full break in

G2D  Amateur Auxiliary; minimizing interference; HF operations

G2E  Digital operating: procedures, procedural signals and common abbreviations

## SUBELEMENT G3 — RADIO WAVE PROPAGATION
### [3 Exam Questions — 3 Groups]

G3A  Sunspots and solar radiation; ionospheric disturbances; propagation forecasting and indices

G3B  Maximum Usable Frequency; Lowest Usable Frequency; propagation

G3C  Ionospheric layers; critical angle and frequency; HF scatter; Near Vertical Incidence Sky waves

## SUBELEMENT G4 — AMATEUR RADIO PRACTICES
### [5 Exam Questions — 5 Groups]

G4A  Station Operation and setup

G4B  Test and monitoring equipment; two-tone test

G4C  Interference with consumer electronics; grounding; DSP

G4D  Speech processors; S meters; sideband operation near band edges

G4E  HF mobile radio installations; emergency and battery powered operation

## SUBELEMENT G5 — ELECTRICAL PRINCIPLES
**[3 Exam Questions — 3 Groups]**

G5A    Reactance; inductance; capacitance; impedance; impedance matching

G5B    The Decibel; current and voltage dividers; electrical power calculations; sine wave root-mean-square (RMS) values; PEP calculations

G5C    Resistors, capacitors and inductors in series and parallel; transformers

## SUBELEMENT G6 — CIRCUIT COMPONENTS
**[3 Exam Questions — 3 Groups]**

G6A    Resistors; capacitors; inductors

G6B    Rectifiers; solid state diodes and transistors; vacuum tubes; batteries

G6C    Analog and digital integrated circuits (IC's); microprocessors; memory; I/O devices; microwave IC's (MMIC's ); display devices

## SUBELEMENT G7 — PRACTICAL CIRCUITS
**[3 Exam Questions — 3 Groups]**

G7A    Power supplies; schematic symbols

G7B    Digital circuits; amplifiers and oscillators

G7C    Receivers and transmitters; filters, oscillators

## SUBELEMENT G8 — SIGNALS AND EMISSIONS
**[2 Exam Questions — 2 Groups]**

G8A    Carriers and modulation: AM; FM; single and double sideband; modulation envelope; overmodulation

G8B    Frequency mixing; multiplication; HF data communications; bandwidths of various modes; deviation

## SUBELEMENT G9 — ANTENNAS AND FEED LINES
**[4 Exam Questions — 4 Groups]**

G9A    Antenna feed lines: characteristic impedance and attenuation; SWR calculation, measurement and effects; matching networks

G9B    Basic antennas

G9C    Directional antennas

G9D    Specialized antennas

## SUBELEMENT G0 — ELECTRICAL AND RF SAFETY
**[2 Exam Questions — 2 Groups]**

G0A    RF safety principles, rules and guidelines; routine station evaluation

G0B    Safety in the ham shack: electrical shock and treatment, safety grounding, fusing, interlocks, wiring, antenna and tower safety

# Commission's Rules

Your General class exam (Element 3) will consist of 35 questions taken from the General class question pool as prepared by the Volunteer Examiner Coordinators' Question Pool Committee. A certain number of questions are taken from each of the 10 subelements. There will be 5 questions from the subelement shown in this chapter. These questions are divided into 5 groups, labeled G1A through G1E.

After most of the explanations in this chapter you will see a reference to Part 97 of the FCC rules set inside square brackets, like [97.301(d)]. This tells you where to look for the exact wording of the rules as they relate to that question. For a complete copy of Part 97, see the ARRL website, **www.arrl.org/part-97-amateur-radio**.

## SUBELEMENT G1 — COMMISSION'S RULES
### [5 Exam Questions — 5 Groups]

### G1A   General Class control operator frequency privileges; primary and secondary allocations

**G1A01**   On which of the following bands is a General Class license holder granted all amateur frequency privileges?

A.  60, 20, 17, and 12 meters
B.  160, 80, 40, and 10 meters
C.  160, 60, 30, 17, 12, and 10 meters
D.  160, 30, 17, 15, 12, and 10 meters

**(C)** These are the bands on which the entire range of mode-restricted segments (such as phone or CW/data) are open to all license classes that have access to the band. Generals, Advanced, and Extra Class licensees all have access to the entire band. [97.301(d), 97.303(s)]

**G1A02**   On which of the following bands is phone operation prohibited?

A.  160 meters
B.  30 meters
C.  17 meters
D.  12 meters

**(B)** The 30 meter band is restricted to CW, RTTY and data transmissions only. [97.305]

**G1A03**   On which of the following bands is image transmission prohibited?

A. 160 meters
B. 30 meters
C. 20 meters
D. 12 meters

**(B)**  The 30 meter band is restricted to CW, RTTY and data transmissions only. Image transmission is also prohibited on the 60 meter band. [97.305]

**G1A04**   Which of the following amateur bands is restricted to communication on only specific channels, rather than frequency ranges?

A. 11 meters
B. 12 meters
C. 30 meters
D. 60 meters

**(D)**  In the US, Amateur Radio is a secondary service to government stations on 60 meters. By limiting amateur operation to specific channels, it is easier for hams to tell when government stations are present and to avoid interfering with them. [97.303 (s)]

**G1A05**   Which of the following frequencies is in the General Class portion of the 40 meter band?

A. 7.250 MHz
B. 7.500 MHz
C. 40.200 MHz
D. 40.500 MHz

**(A)**  General Class licensees have access to the following portions of the 40 meter band (f = 300 / 40 = 7.5 MHz): 7.025 - 7.125 MHz on CW/RTTY/Data and from 7.175 - 7.300 MHz on CW/Phone/Image. [97.301(d)]

**G1A06**   Which of the following frequencies is in the 12 meter band?

A. 3.940 MHz
B. 12.940 MHz
C. 17.940 MHz
D. 24.940 MHz

**(D)**  Use the formula f = 300 / 12 = 25 MHz to get the approximate frequency range of the 12 meter band. [97.301(d)]

**G1A07** Which of the following frequencies is within the General class portion of the 75 meter phone band?

   A. 1875 kHz
   B. 3750 kHz
   C. 3900 kHz
   D. 4005 kHz

**(C)** Although the 75 and 80 meter bands are part of a single amateur band, the difference in wavelength is enough for amateurs to make a distinction between 75 meters at the higher frequencies and 80 meters at the lower frequencies. General Class licensees have access to the following portions of the 75 meter band (f = 300 / 75 = 4.0 MHz): 3.800 - 4.000 MHz on CW/Phone/Image. [97.301(d)]

**G1A08** Which of the following frequencies is within the General Class portion of the 20 meter phone band?

   A. 14005 kHz
   B. 14105 kHz
   C. 14305 kHz
   D. 14405 kHz

**(C)** General Class licensees have access to the following portions of the 20 meter band (f = 300 / 20 = 15 MHz): 14.025 - 14.150 MHz on CW/RTTY/Data and from 14.225 - 14.350 MHz on CW/Phone/Image. [97.301(d)]

**G1A09** Which of the following frequencies is within the General Class portion of the 80 meter band?

   A. 1855 kHz
   B. 2560 kHz
   C. 3560 kHz
   D. 3650 kHz

**(C)** See G1A07. General Class licensees have access to the following portions of the 80 meter band (f = 300 / 80 = 3.75 MHz): 3.525 - 3.600 MHz on CW/RTTY/Data. [97.301(d)]

**G1A10** Which of the following frequencies is within the General Class portion of the 15 meter band?

   A. 14250 kHz
   B. 18155 kHz
   C. 21300 kHz
   D. 24900 kHz

**(C)** General Class licensees have access to the following portions of the 15 meter band (f = 300 / 15 = 20 MHz): 21.025 - 21.200 MHz on CW/RTTY/Data and from 21.275 - 21.450 MHz on CW/Phone/Image. [97.301(d)]

**G1A11**    Which of the following frequencies is available to a control operator holding a General Class license?

A.  28.020 MHz
B.  28.350 MHz
C.  28.550 MHz
D.  All of these choices are correct

**(D)** The 28 MHz (10 meter) band is one of the bands on which the entire range of mode-restricted segments (such as phone or CW/data) are open to all license classes that have access to the band. Generals, Advanced, and Extra Class licensees all have access to the entire band. [97.301(d)]

**G1A12**    When General Class licensees are not permitted to use the entire voice portion of a particular band, which portion of the voice segment is generally available to them?

A.  The lower frequency end
B.  The upper frequency end
C.  The lower frequency end on frequencies below 7.3 MHz and the upper end on frequencies above 14.150 MHz
D.  The upper frequency end on frequencies below 7.3 MHz and the lower end on frequencies above 14.150 MHz

**(B)** If you look at the US Amateur Band chart available in the *General Class License Manual* or on the ARRL website at **www.arrl.org/files/file/Hambands_color.pdf** you will see that in the bands on which there are mode-restricted segments, such as 80 meters, General Class licensees have access to the higher frequencies of the segment. [97.301]

**G1A13**    Which, if any, amateur band is shared with the Citizens Radio Service?

A.  10 meters
B.  12 meters
C.  15 meters
D.  None

**(D)** Citizens Band operators may not operate on amateur bands or on any frequency not assigned to the Citizens Radio Service. [97.303]

**G1A14** Which of the following applies when the FCC rules designate the Amateur Service as a secondary user on a band?

A. Amateur stations must record the call sign of the primary service station before operating on a frequency assigned to that station
B. Amateur stations are allowed to use the band only during emergencies
C. Amateur stations are allowed to use the band only if they do not cause harmful interference to primary users
D. Amateur stations may only operate during specific hours of the day, while primary users are permitted 24 hour use of the band

**(C)** You should always listen before you transmit. This is especially important on bands where Amateur Radio is a secondary service, such as the 30 or 60 meter band. Amateurs are only permitted to use these frequencies if they do not cause harmful interference to the primary users. If you hear a station in the primary service or receive interference from such a station you should immediately change frequencies. Otherwise you might be causing interference to the primary station. [97.303]

**G1A15** What is the appropriate action if, when operating on either the 30 or 60 meter bands, a station in the primary service interferes with your contact?

A. Notify the FCC's regional Engineer in Charge of the interference
B. Increase your transmitter's power to overcome the interference
C. Attempt to contact the station and request that it stop the interference
D. Move to a clear frequency

**(D)** See G1A14. [97.303]

---

**G1B** **Antenna structure limitations; good engineering and good amateur practice; beacon operation; restricted operation; retransmitting radio signals**

**G1B01** What is the maximum height above ground to which an antenna structure may be erected without requiring notification to the FAA and registration with the FCC, provided it is not at or near a public-use airport?

A. 50 feet
B. 100 feet
C. 200 feet
D. 300 feet

**(C)** FCC regulations require approval if your antenna would be more than 200 feet above ground level at its site. This includes the antenna, the supports and anything else attached to the structure. (Additional FCC restrictions apply if the antenna is within about 4 miles of a public use airport or heliport.) [97.15(a)]

**G1B02** With which of the following conditions must beacon stations comply?

A. A beacon station may not use automatic control
B. The frequency must be coordinated with the National Beacon Organization
C. The frequency must be posted on the Internet or published in a national periodical
D. There must be no more than one beacon signal in the same band from a single location

**(D)** A beacon station normally transmits a signal for operators to observe propagation and reception characteristics. For this purpose, FCC rules specifically allow an amateur beacon to transmit one-way communications. [97.203(b)]

**G1B03** Which of the following is a purpose of a beacon station as identified in the FCC Rules?

A. Observation of propagation and reception, or other related activities
B. Automatic identification of repeaters
C. Transmission of bulletins of general interest to Amateur Radio licensees
D. Identifying net frequencies

**(A)** See G1B02. [97.3(a)(9)]

**G1B04** Which of the following must be true before amateur stations may provide communications to broadcasters for dissemination to the public?

A. The communications must directly relate to the immediate safety of human life or protection of property and there must be no other means of communication reasonably available before or at the time of the event
B. The communications must be approved by a local emergency preparedness official and conducted on officially designated frequencies
C. The FCC must have declared a state of emergency
D. All of these choices are correct

**(A)** Amateurs are not allowed to be involved with any activity related to program production or news gathering for broadcasting to the general public unless it is directly related to an emergency involving an immediate life or property-threatening situation and there is no other method by which the information can be transmitted. If there is an alternative communication system available, even if it is slower, the amateur station may not transmit the information for a news broadcast or related program production. [97.113(b)]

**G1B05** When may music be transmitted by an amateur station?

A. At any time, as long as it produces no spurious emissions
B. When it is unintentionally transmitted from the background at the transmitter
C. When it is transmitted on frequencies above 1215 MHz
D. When it is an incidental part of a space shuttle or ISS retransmission

**(D)** Normally, music may not be transmitted by an amateur station. This is to avoid infringing upon commercial broadcast activities. Music is often used in communications between ground control and the space shuttle or International Space Station (ISS) for such things as waking up the astronauts in the morning. In this case, since it is an incidental part of the transmission and not the primary purpose of it, the music can be retransmitted by the amateur station along with the rest of the space shuttle transmission, as long as the amateur station has NASA's permission to retransmit the Shuttle (or ISS) audio. [97.113(a)(4), (e)]

**G1B06** When is an amateur station permitted to transmit secret codes?

A. During a declared communications emergency
B. To control a space station
C. Only when the information is of a routine, personal nature
D. Only with Special Temporary Authorization from the FCC

**(B)** An amateur station may never transmit in such a manner as to obscure the meaning of two-way communication. The use of standard abbreviations does not violate this rule, since their meaning is well known. The exception is space telecommand operations where the commands and data may be coded. (This helps prevent unauthorized stations from transmitting telecommand messages to the spacecraft.) When controlling a satellite from a ground station, the transmissions may consist of specially coded messages intended to facilitate communications or related to the function of the spacecraft. Telecommand is not two-way communication, however; it is one-way communication. [97.113(a)(4) and 97.207(f)]

**G1B07** What are the restrictions on the use of abbreviations or procedural signals in the Amateur Service?

A. Only "Q" codes are permitted
B. They may be used if they do not obscure the meaning of a message
C. They are not permitted
D. Only "10-codes" are permitted

**(B)** The use of common abbreviations and procedural signals is standard practice and does not obscure the meaning of a message because their meaning is well known. Any use of abbreviations or codes for the purpose of obscuring the meaning of a communication is prohibited. (see G1B06) [97.113(a)(4)]

**G1B08**  When choosing a transmitting frequency, what should you do to comply with good amateur practice?

A. Review FCC Part 97 Rules regarding permitted frequencies and emissions

B. Follow generally accepted band plans agreed to by the Amateur Radio community.

C. Before transmitting, listen to avoid interfering with ongoing communication

D. All of these choices are correct

**(D)**  Choosing a frequency is straightforward: Be sure the frequency is authorized to General class licensees, follow the band plan under normal circumstances, and listen to the frequency to avoid interfering with ongoing communications.

**G1B09**  When may an amateur station transmit communications in which the licensee or control operator has a pecuniary (monetary) interest?

A. When other amateurs are being notified of the sale of apparatus normally used in an amateur station and such activity is not done on a regular basis

B. Only when there is no other means of communications readily available

C. When other amateurs are being notified of the sale of any item with a monetary value less than $200 and such activity is not done on a regular basis

D. Never

**(A)**  In general, amateurs are forbidden to receive any kind of compensation, financial or otherwise, for conducting communications on amateur frequencies. Amateurs are, however, permitted to conduct a limited amount of personal business, such as participating in "swap-and-shop" nets conducted on amateur frequencies for local amateurs to buy, sell and trade amateur equipment. [97.113(a)(3)]

**G1B10**  What is the power limit for beacon stations?

A. 10 watts PEP output

B. 20 watts PEP output

C. 100 watts PEP output

D. 200 watts PEP output

**(C)**  100 watts of output power is a good compromise, enabling a beacon station to transmit a signal strong enough to be heard when propagation isn't the best. Similarly, when propagation is good, a 100-watt signal is not so strong as to cause interference to stations on nearby frequencies. [97.203(c)]

**G1B11**  How does the FCC require an amateur station to be operated in all respects not specifically covered by the Part 97 rules?

A. In conformance with the rules of the IARU
B. In conformance with Amateur Radio custom
C. In conformance with good engineering and good amateur practice
D. All of these answers are correct

**(C)** The FCC Rules grant amateurs a lot freedom in the ways they choose to operate, more than in any other service. It is impossible for the service's broad rules and regulations to cover every situation that might possibly arise. You are expected to use common sense in those situations where an exact rule does not apply. Your station and its operation should always follow good engineering design and good amateur practice. [97.101(a)]

**G1B12**  Who or what determines "good engineering and good amateur practice" as applied to operation of an amateur station in all respects not covered by the Part 97 rules?

A. The FCC
B. The Control Operator
C. The IEEE
D. The ITU

**(A)** The FCC does not publish a list of what constitutes "good engineering and good amateur practice" because the state of the radio art is continually improving. Nevertheless, when questions arise, the FCC is the agency that determines what standards should be applied. [97.101(a)]

## G1C  Transmitter power regulations; data emission standards

**G1C01**  What is the maximum transmitting power an amateur station may use on 10.140 MHz?

A. 200 watts PEP output
B. 1000 watts PEP output
C. 1500 watts PEP output
D. 2000 watts PEP output

**(A)** The general rule is that maximum power is limited to 1500 watts PEP output, although there are exceptions where less power is allowed. One such exception is the 30 meter band, 10.100 - 10.150 MHz, where the maximum power for US hams is 200 watts. (These frequencies are just above the short-wave time broadcasts of WWV and WWVH at 10.0 MHz.) [97.313(c)(1)]

**G1C02**  What is the maximum transmitting power an amateur station may use on the 12 meter band?

A. 1500 watts PEP output, except for 200 watts PEP output in the Novice portion
B. 200 watts PEP output
C. 1500 watts PEP output
D. An effective radiated power equivalent to 50 watts from a half wave dipole

**(C)** The maximum power allowed is 1500 watts PEP output from 24.890 - 24.990 MHz. See also G1C01. [97.313(a), (b)]

**G1C03**  What is the maximum bandwidth permitted by FCC rules for amateur radio stations when transmitting on USB frequencies in the 60 meter band?

A. 2.8 kHz
B. 5.6 kHz
C. 1.8 kHz
D. 3 kHz

**(A)** The FCC Rules for operating on the amateur 60 meter band tell us that "Amateur stations must ensure that their transmission occupies only the 2.8 kHz centered around each" of the operating channels. So the maximum transmitted bandwidth of your upper sideband signal (the only operating mode allowed on this band) is 2.8 kHz. A properly adjusted SSB transmitter normally has a bandwidth of 2.5 to 2.8 kHz. [97.303(s)]

**G1C04**  Which of the following is a limitation on transmitter power in the 14 MHz band?

A. Only the minimum power necessary to carry out the desired communications should be used
B. Power must be limited to 200 watts when transmitting between 14.100 MHz and 14.150 MHz
C. Power should be limited as necessary to avoid interference to another radio service on the frequency
D. Effective radiated power cannot exceed 3000 watts

**(A)** Although the maximum power allowed is 1500 watts PEP output, amateurs should use only the power level needed to carry out communications. [97.313(a)]

**G1C05**  Which of the following is a limitation on transmitter power in the 28 MHz band?

A. 100 watts PEP output
B. 1000 watts PEP output
C. 1500 watts PEP output
D. 2000 watts PEP output

**(C)** The maximum power allowed is 1500 watts PEP output on the entire band. See also G1C01. [97.313(b)]

**G1C06** Which of the following is a limitation on transmitter power in the 1.8 MHz band?

A. 200 watts PEP output
B. 1000 watts PEP output
C. 1200 watts PEP output
D. 1500 watts PEP output

**(D)** The maximum power allowed is 1500 watts PEP output on the entire band. See also G1C01. [97.313(b)]

**G1C07** What is the maximum symbol rate permitted for RTTY or data emission transmission on the 20 meter band?

A. 56 kilobaud
B. 19.6 kilobaud
C. 1200 baud
D. 300 baud

**(D)** The symbol rate of digital signals is restricted to make sure they do not consume too much bandwidth at the expense of other modes. **Table G1-1** shows the limits by band. [97.305(c), 97.307(f)(3)]

**Table G1-1**
**Maximum Symbol Rates and Bandwidth**

| Band | Symbol Rate (baud) | Bandwidth (kHz) |
|---|---|---|
| 160 through 10 m | 300 | 1 |
| 10 m | 1200 | 1 |
| 6 m, 2 m | 19.6k | 20 |
| 1.25 m, 70 cm | 56k | 100 |
| 33 cm and above | no limit | no limit |

**G1C08** What is the maximum symbol rate permitted for RTTY or data emission transmitted at frequenciesbelow 28 MHz?

A. 56 kilobaud
B. 19.6 kilobaud
C. 1200 baud
D. 300 baud

**(D)** See G1C07. [97.305(c) and 97.307(f)(3)]

**G1C09** What is the maximum symbol rate permitted for RTTY or data emission transmitted on the 1.25 meter and 70 centimeter bands?

A. 56 kilobaud
B. 19.6 kilobaud
C. 1200 baud
D. 300 baud

**(A)** See G1C07. [97.305(c) and 97.307(f)(5)]

**G1C10** What is the maximum symbol rate permitted for RTTY or data emission transmissions on the 10 meter band?

A. 56 kilobaud
B. 19.6 kilobaud
C. 1200 baud
D. 300 baud

**(C)** See G1C07. [97.305(c) and 97.307(f)(4)]

**G1C11** What is the maximum symbol rate permitted for RTTY or data emission transmissions on the 2 meter band?

A. 56 kilobaud
B. 19.6 kilobaud
C. 1200 baud
D. 300 baud

**(B)** See G1C07. [97.305(c) and 97.307(f)(5)]

## G1D    Volunteer Examiners and Volunteer Examiner Coordinators; temporary identification

**G1D01** Which of the following is a proper way to identify when transmitting using phone on General Class frequencies if you have a CSCE for the required elements but your upgrade from Technician has not appeared in the FCC database?

A. Give your call sign followed by the words "General Class"
B. No special identification is needed
C. Give your call sign followed by "slant AG"
D. Give your call sign followed by the abbreviation "CSCE"

**(C)** You must add a "temporary identifier" to your call sign so that stations receiving your transmissions can verify that you are authorized to transmit on that frequency. If a temporary identifier was not used between the time you pass your exam and the time at which your new privileges appear in the FCC database, it would appear that you were transmitting on a frequency for which you were not authorized. When you upgrade to Extra class, you'll append "temporary AE" to your call sign. [97.119(f)(2)]

**G1D02** What license examinations may you administer when you are an accredited VE holding a General Class operator license?

A. General and Technician
B. General only
C. Technician only
D. Extra, General and Technician

**(C)** Holders of a General class operator license may only administer examinations for license levels below theirs: Technician. General class licensees may participate as a VE in any exam session, but they may not be the primary VEs administering the General or Extra class exams. [97.507(b)(3)(i)]

**G1D03** On which of the following band segments may you operate on if you are a Technician Class operator and have a CSCE for General Class privileges?

A. Only the Technician band segments until your upgrade is posted on the FCC database

B. Only on the Technician band segments until your license arrives in the mail

C. On any General or Technician Class band segment

D. On any General or Technician Class band segment except 30 and 60 meters

**(C)** You may begin using the General class privileges immediately on receiving your CSCE, but you must append the temporary identifier to your call sign as described in the discussion for G1D01. [97.9(b)]

**G1D04** Which of the following is a requirement for administering a Technician Class operator examination?

A. At least three VEC-accredited General Class or higher VEs must be present

B. At least two VEC-accredited General Class or higher VEs must be present

C. At least two General Class or higher VEs must be present, but only one need be VEC accredited

D. At least three VEs of Technician Class or higher must be present

**(A)** All license exams are administered through the Volunteer Examiner Coordinator (VEC) system. VEs (Volunteer Examiners) must be accredited by a VEC. There must be three VEC-accredited VEs present at every exam session. Technician exams are administered by General class or higher VEs. [97.509(a), (b)]

**G1D05** Which of the following is sufficient for you to be an administering VE for a Technician Class operator license examination?

A. Notification to the FCC that you want to give an examination

B. Receipt of a CSCE for General class

C. Possession of properly obtained telegraphy license

D. A FCC General class or higher license and VEC accreditation

**(D)** All license exams are administered through the Volunteer Examiner Coordinator (VEC) system. VEs (Volunteer Examiners) must be accredited by a VEC. There must be three VEC-accredited VEs present at every exam session. Technician exams are administered by General class or higher VEs. As soon as the FCC issues your General class license (meaning that it has appeared in the FCC database) and you receive your accreditation from a VEC, you can participate in exam sessions for Technician license exams. [97.509(b)(3)(i) and 97.509(b)(1)]

**G1D06** When must you add the special identifier "AG" after your call sign if you are a Technician Class licensee and have a CSCE for General Class operator privileges, but the FCC has not yet posted your upgrade on its web site?

A. Whenever you operate using General Class frequency privileges
B. Whenever you operate on any amateur frequency
C. Whenever you operate using Technician frequency privileges
D. A special identifier is not required as long as your General Class license application has been filed with the FCC

**(A)** See G1D01. [97.119(f)(2)]

**G1D07** Volunteer Examiners are accredited by what organization?

A. The Federal Communications Commission
B. The Universal Licensing System
C. A Volunteer Examiner Coordinator
D. The Wireless Telecommunications Bureau

**(C)** A Volunteer Examiner Coordinator (VEC) organization is responsible for certifying Volunteer Examiners and evaluating the results of all exam sessions administered by them. VECs also process all of the license application paperwork and submit it to the FCC. [97.509(b)(1)]

**G1D08** What criteria must be met for a non-U.S. citizen to be an accredited Volunteer Examiner?

A. The person must be a resident of the U.S. for a minimum of 5 years
B. The person must hold an FCC-granted Amateur Radio license of General class or above
C. The person's home citizenship must be in the ITU 2 region
D. None of these choices is correct; non U.S. citizens cannot be volunteer examiners

**(B)** A VE's citizenship does not matter, only whether the individual has demonstrated adequate knowledge of the US Amateur Service rules by passing the appropriate license exams. [97.509 (b)(3)]

**G1D09** How long is a Certificate of Successful Completion of Examination (CSCE) valid for exam element credit?

A. 30 days
B. 180 days
C. 365 days
D. For as long as your current license is valid

**(C)** Although your new license class should appear in the FCC database within a few days of passing your examination, should there be a delay, remember that the CSCE is only good for 365 days. After that time, you'll have to re-take the examination! [97.9(b)]

**G1D10** What is the minimum age that one must be to qualify as an accredited Volunteer Examiner?

A. 12 years
B. 18 years
C. 21 years
D. There is no age limit

**(B)** 18 years old was determined to be an appropriate age to properly manage an amateur examination session. [97.509(b)(2)]

## G1E Control categories; repeater regulations; harmful interference; third party rules; ITU regions

**G1E01** Which of the following would disqualify a third party from participating in stating a message over an amateur station?

A. The third party's amateur license has ever been revoked
B. The third party is not a U.S. citizen
C. The third party is a licensed amateur
D. The third party is speaking in a language other than English, French, or Spanish

**(A)** Third-party communication is available to anyone except someone with a revoked amateur license from any country. This prevents someone whose ability to make use of Amateur Radio was taken away from regaining access to amateur frequencies under the guise of third party communications. [97.115(b)(2)]

**G1E02** When may a 10 meter repeater retransmit the 2 meter signal from a station having a Technician Class control operator?

A. Under no circumstances
B. Only if the station on 10 meters is operating under a Special Temporary Authorization allowing such retransmission
C. Only during an FCC declared general state of communications emergency
D. Only if the 10 meter repeater control operator holds at least a General Class license

**(D)** FCC rules allow any holder of an amateur license to be the control operator of a repeater. The control operator of the repeater must have privileges on the frequency on which the repeater is transmitting, however. A 10 meter repeater must have a General class or higher control operator because Technician and Novice licensees don't have privileges on the 10 meter repeater band. A 10 meter repeater may retransmit the 2 meter signal from a Technician class operator because the 10 meter control operator holds at least a General class license. (Of course the operator transmitting to the repeater must have privileges on the frequency on which he or she is transmitting.) [97.205(a)]

**G1E03** In what ITU region is operation in the 7.175 to 7.300 MHz band permitted for a control operator holding an FCC-issued General Class license?

A. Region 1
B. Region 2
C. Region 3
D. All three regions

**(B)** Amateur allocations vary between the different ITU regions of the world. FCC Rule 97.301 contains a complete listing of allocations for US hams. Parts (a) and (d) of that section contain the Region 2 frequency allocations that apply to General class amateurs operating from the US. [97.301]

**G1E04** Which of the following conditions require an Amateur Radio station licensee to take specific steps to avoid harmful interference to other users or facilities?

A. When operating within one mile of an FCC Monitoring Station
B. When using a band where the Amateur Service is secondary
C. When a station is transmitting spread spectrum emissions
D. All of these choices are correct

**(D)** Aside from the general requirement to avoid causing harmful interference to other licensed stations and primary service licensees, there are several specific instances in which amateurs must take extra steps to avoid interference. FCC Monitoring Stations require an environment free of strong or spurious signals that can cause interference. The location of monitoring stations can be determined from a regional FCC office. Spread spectrum (SS) transmissions, because of their nature, have the potential to interfere with fixed frequency stations, so SS users should be sure their transmissions will not cause interference. [97.13(b), 97.311(b), 97.303]

**G1E05** What types of messages for a third party in another country may be transmitted by an amateur station?

A. Any message, as long as the amateur operator is not paid
B. Only messages for other licensed amateurs
C. Only messages relating to Amateur Radio or remarks of a personal character, or messages relating to emergencies or disaster relief
D. Any messages, as long as the text of the message is recorded in the station log

**(C)** The FCC and other licensing authorities want to be very sure that the amateur service is not abused to provide communications that should properly be conducted through commercial or government services. As a result, third-party communication is restricted to the types of messages in answer C. [97.115(a)(2), 97.117]

**G1E06** Which of the following applies in the event of interference between a coordinated repeater and an uncoordinated repeater?

A. The licensee of the non-coordinated repeater has primary responsibility to resolve the interference
B. The licensee of the coordinated repeater has primary responsibility to resolve the interference
C. Both repeater licensees share equal responsibility to resolve the interference
D. The frequency coordinator bears primary responsibility to resolve the interference

**(A)** The FCC considers repeater frequency coordination to be "good engineering and amateur practice." As such, amateurs are expected to use frequency coordination methods whenever the potential for interference exists. As a consequence, the burden of resolving interference between a coordinated and non-coordinated repeater system falls on the operator of the non-coordinated system. [97.205(c)]

**G1E07** With which foreign countries is third-party traffic prohibited, except for messages directly involving emergencies or disaster relief communications?

A. Countries in ITU Region 2
B. Countries in ITU Region 1
C. Every foreign country, unless there is a third-party agreement in effect with that country
D. Any country which is not a member of the International Amateur Radio Union (IARU)

**(C)** The general rule is that third-party traffic with amateurs in any country outside the US is prohibited unless specifically permitted by a third-party agreement between the United States and that country. Don't assume that it is permitted. Check **Table G1-2** which lists countries having third-party agreements with the United States. If the country is not listed, you may not exchange third-party traffic with amateurs operating in that country. [97.115(a)(2)]

## Table G1-1

### Third-Party Traffic Agreements List

Occasionally, DX stations may ask you to pass a third-party message to a friend or relative in the States. This is all right as long as the US has signed an official third-party traffic agreement with that particular country, or the third party is a licensed amateur. The traffic must be noncommercial and of a personal, unimportant nature. During an emergency, the US State Department will often work out a special temporary agreement with the country involved. But in normal times, never handle traffic without first making sure it is legally permitted.

*US Amateurs May Handle Third-Party Traffic With:*

| | | | |
|---|---|---|---|
| C5 | The Gambia | TI | Costa Rica |
| CE | Chile | T9 | Bosnia-Herzegovina |
| CO | Cuba | V2 | Antigua and Barbuda |
| CP | Bolivia | V3 | Belize |
| CX | Uruguay | V4 | St Kitts and Nevis |
| D6 | Federal Islamic Rep. | V6 | Federated States of |
| | of the Comoros | | Micronesia |
| DU | Philippines | V7 | Marshall Islands |
| EL | Liberia | VE | Canada |
| GB | United Kingdom | VK | Australia |
| HC | Ecuador | VR6 | Pitcairn Island* |
| HH | Haiti | XE | Mexico |
| HI | Dominican Republic | YN | Nicaragua |
| HK | Colombia | YS | El Salvador |
| HP | Panama | YV | Venezuela |
| HR | Honduras | ZP | Paraguay |
| J3 | Grenada | ZS | South Africa |
| J6 | St Lucia | 3DA | Swaziland |
| J7 | Dominica | 4U1ITU | ITU - Geneva |
| J8 | St Vincent and the | 4U1VIC | VIC - Vienna |
| | Grenadines | 4X | Israel |
| JY | Jordan | 6Y | Jamaica |
| LU | Argentina | 8R | Guyana |
| OA | Peru | 9G | Ghana |
| PY | Brazil | 9L | Sierra Leone |
| TA | Turkey | 9Y | Trinidad and Tobago |
| TG | Guatemala | | |

*Notes:*

*Since 1970, there has been an informal agreement between the United Kingdom and the US, permitting Pitcairn and US amateurs to exchange messages concerning medical emergencies, urgent need for equipment or supplies, and private or personal matters of island residents.

Please note that Region 2 of the International Amateur Radio Union (IARU) has recommended that international traffic on the 20 and 15-meter bands be conducted on 14.100-14.150, 14.250-14.350, 21.150-21.200 and 21.300-21.450 MHz. The IARU is the alliance of Amateur Radio societies from around the world; Region 2 comprises member-societies in North, South and Central America and the Caribbean.

At the end of an exchange of third-party traffic with a station located in a foreign country, an FCC-licensed amateur must transmit the call sign of the foreign station as well as his own call sign.

Current as of January 2011; see **www.arrl.org/third-party-operating-agreements** for the latest information.

**G1E08** Which of the following is a requirement for a non-licensed person to communicate with a foreign Amateur Radio station from a station with an FCC-granted license at which a licensed control operator is present?

A. Information must be exchanged in English
B. The foreign amateur station must be in a country with which the United States has a third party agreement
C. The control operator must have at least a General Class license
D. All of these choices are correct

**(B)** The non-licensed person is, by definition, a third-party and any messages you allow them to send or send on their behalf are third-party traffic. The non-licensed person cannot have had an amateur license revoked as discussed in question G1E01. [97.115(a) and (b)]

**G1E09** What language must you use when identifying your station if you are using a language other than English in making a contact using phone emission?

A. The language being used for the contact
B. Any language if the US has a third party agreement with that country
C. English
D. Any language of a country that is a member of the ITU

**(C)** Identification by all US-licensed stations when using phone transmissions must be performed in English. [97.119(b)(2)]

**G1E10** What portion of the 10 meter band is available for repeater use?

A. The entire band
B. The portion between 28.1 MHz and 28.2 MHz
C. The portion between 28.3 MHz and 28.5 MHz
D. The portion above 29.5 MHz

**(D)** The only HF authorization for repeater stations is on the 10 meter band between 29.5 and 29.7 MHz. [97.205 (b)]

# Operating Procedures

Your General class exam (Element 3) will consist of 35 questions taken from the General class question pool as prepared by the Volunteer Examiner Coordinators' Question Pool Committee. A certain number of questions are taken from each of the 10 subelements. There will be 5 questions from the subelement shown in this chapter.

These questions are divided into 5 groups, labeled G2A through G2E. After some of the explanations in this chapter you will see a reference to Part 97 of the FCC rules set inside square brackets, like [97.301(d)]. This tells you where to look for the exact wording of the rules as they relate to that question. For a complete copy of Part 97, see the ARRL website, **www.arrl.org/part-97-amateur-radio**.

## SUBELEMENT G2 — OPERATING PROCEDURES
[5 Exam Questions — 5 Groups]

**G2A    Phone operating procedures; USB/LSB utilization conventions; procedural signals; breaking into a QSO in progress; VOX operation**

**G2A01**    Which sideband is most commonly used for voice communications on frequencies of 14 MHz or higher?
A.  Upper sideband
B.  Lower sideband
C.  Vestigial sideband
D.  Double sideband

**(A)** Single-sideband (SSB) modulation removes the carrier and one sideband from an AM signal to conserve spectrum and for improved power efficiency. Amateurs normally use the upper sideband for 20 meter phone operation. Whether the upper or lower sideband is used is strictly a matter of convention and not of regulation. The convention to use the lower sideband on the bands below 9 MHz and the upper sideband on the higher-frequency bands developed from the design requirements of early SSB transmitters. Although modern amateur SSB equipment is more flexible, the convention persists. If everyone else on a particular band is using a certain sideband, you will need to use the same one in order to be able to communicate.

**G2A02**  Which of the following modes is most commonly used for voice communications on the 160, 75, and 40 meter bands?

A. Upper sideband
B. Lower sideband
C. Vestigial sideband
D. Double sideband

**(B)**  Amateurs normally use the lower sideband for 160, 75 and 40 meter phone operation. (see the discussion for question G2A02)

**G2A03**  Which of the following is most commonly used for SSB voice communications in the VHF and UHF bands?

A. Upper sideband
B. Lower sideband
C. Vestigial sideband
D. Double sideband

**(A)**  Amateurs normally use the upper sideband for VHF and UHF phone operation. (see the discussion for question G2A02)

**G2A04**  Which mode is most commonly used for voice communications on the 17 and 12 meter bands?

A. Upper sideband
B. Lower sideband
C. Vestigial sideband
D. Double sideband

**(A)**  Amateurs normally use the upper sideband for 17 and 12 meter phone operation. (see the discussion for question G2A02)

**G2A05**  Which mode of voice communication is most commonly used on the high frequency amateur bands?

A. Frequency modulation
B. Double sideband
C. Single sideband
D. Phase modulation

**(C)**  Most amateurs who use voice communications on the high frequency bands use single sideband (SSB) voice. There are some operators who prefer the high-fidelity audio of double-sideband full-carrier amplitude modulation (AM). AM requires more than twice the bandwidth of an SSB signal, however. There is also some frequency modulated (FM) and phase modulated (PM) voice operation on the 10 meter band, but that mode also requires a much wider bandwidth than SSB. Some amateurs are beginning to experiment with digitally encoded voice communications, but SSB is the most common HF voice mode.

**G2A06** Which of the following is an advantage when using single sideband as compared to other analog voice modes on the HF amateur bands?

A. Very high fidelity voice modulation
B. Less bandwidth used and higher power efficiency
C. Ease of tuning on receive and immunity to impulse noise
D. Less subject to static crashes (atmospherics)

**(B)** Single sideband (SSB) voice communication is used much more frequently than other voice modes on the HF bands because it uses less spectrum space. The RF carrier is not transmitted with an SSB signal. That means SSB transmissions are more power efficient, since the full transmitter power can be used to transmit the one sideband rather than being divided between the two sidebands and the carrier as it would be for AM.

**G2A07** Which of the following statements is true of the single sideband (SSB) voice mode?

A. Only one sideband and the carrier are transmitted; the other sideband is suppressed
B. Only one sideband is transmitted; the other sideband and carrier are suppressed
C. SSB voice transmissions have higher average power than any other mode
D. SSB is the only mode that is authorized on the 160, 75 and 40 meter amateur bands

**(B)** Single sideband (SSB) voice transmissions are identified by which sideband is used. If the sideband with a frequency lower than the RF carrier frequency is used, then the signal is known as a lower sideband (LSB) transmission. If the sideband with a frequency higher than the RF carrier frequency is used, then the signal is known as an upper sideband (USB) transmission. In both cases the opposite sideband is suppressed. Amateurs normally use lower sideband on the 160, 75/80 and 40 meter bands, and upper sideband on the 20, 17, 15, 12 and 10 meter bands. This is not a requirement of the FCC Rules in Part 97, though. It is simply by common agreement. FCC Rules do, however, require amateurs to use USB on the five channels of the 60 meter band.

**G2A08** Which of the following is a recommended way to break into a conversation when using phone?

A. Say "QRZ" several times followed by your call sign
B. Say your call sign during a break between transmissions from the other stations
C. Say "Break. Break. Break." and wait for a response
D. Say "CQ" followed by the call sign of either station

**(B)** To break into a conversation, you will have to wait until both stations are listening so that your signal will be heard. In order that your transmissions be identified during this brief period, simply state your call sign. No "over" or "break" is required, nor do you have to give either of the transmitting station's call signs.

**G2A09** Why do most amateur stations use lower sideband on the 160, 75 and 40 meter bands?

A. Lower sideband is more efficient than upper sideband at these frequencies
B. Lower sideband is the only sideband legal on these frequency bands
C. Because it is fully compatible with an AM detector
D. Current amateur practice is to use lower sideband on these frequency bands

**(D)** There are no FCC Part 97 Rules about which sideband is used on any band except on 60 meters. Amateurs have adopted the practice of using lower sideband (LSB) on the 160, 75/80 and 40 meter bands, and upper sideband (USB) on the 20, 17, 15, 12 and 10 meter bands. This practice began in the early days of SSB operation, because of some design requirements of early SSB transmitters.

**G2A10** Which of the following statements is true of SSB VOX operation?

A. The received signal is more natural sounding
B. VOX allows "hands free" operation
C. Frequency spectrum is conserved
D. Provides more power output

**(B)** The purpose of a voice operated transmit (VOX) circuit is to provide automatic transmit/receive (TR) switching within an amateur station. By simply speaking into the microphone, the antenna is connected to transmitter, the receiver is muted and the transmitter is activated. When you stop speaking, the VOX circuit switches everything back to receive. Using VOX allows hands-free operation.

**G2A11** What does the expression "CQ DX" usually indicate?

A. A general call for any station
B. The caller is listening for a station in Germany
C. The caller is looking for any station outside their own country
D. A distress call

**(C)** DX means "distant stations" in ham jargon, so combining CQ which means "I am calling any station" with DX means "I am calling any distant station." You may also hear stations making targeted calls by combining CQ with some other description, such as "CQ mobile stations" or "CQ California." It is polite to avoid responding if you are not of the type of station being called.

## G2B    Operating courtesy; band plans, emergencies, including drills and emergency communications

**G2B01**    Which of the following is true concerning access to frequencies?

A. Nets always have priority
B. QSO's in process always have priority
C. No one has priority access to frequencies, common courtesy should be a guide
D. Contest operations must always yield to non-contest use of frequencies

**(C)** Except when the FCC has declared there to be a communications emergency and designated specific frequencies for emergency communications, no single or group of amateurs has priority on any amateur frequency. Good operating practice is to use the flexibility of the amateur service to avoid interference and minimize any interference from your operation.

**G2B02**    What is the first thing you should do if you are communicating with another amateur station and hear a station in distress break in?

A. Continue your communication because you were on frequency first
B. Acknowledge the station in distress and determine what assistance may be needed
C. Change to a different frequency
D. Immediately cease all transmissions

**(B)** Whenever you hear a station in distress (where there is immediate threat to human life or property), you should take whatever action is necessary to determine what assistance that station needs and attempt to provide it. Don't assume that some other station will handle the emergency; you may be the only station receiving the distress signal. If you do hear a station in distress, the first thing you should do is to acknowledge that you heard the station, and then ask the operator where they are located and what assistance they need.

**G2B03**    If propagation changes during your contact and you notice increasing interference from other activity on the same frequency, what should you do?

A. Tell the interfering stations to change frequency
B. Report the interference to your local Amateur Auxiliary Coordinator
C. As a common courtesy, move your contact to another frequency
D. Increase power to overcome interference

**(C)** Good operating practice suggests that whoever can most easily resolve an interference problem be the one to do so. If you begin to have interference from other activity on the same frequency, moving your contact to another frequency may be the simplest thing to do. Switching antennas or rotating a beam antenna may also achieve the same results.

**G2B04** When selecting a CW transmitting frequency, what minimum frequency separation should you allow in order to minimize interference to stations on adjacent frequencies?

A. 5 to 50 Hz
B. 150 to 500 Hz
C. 1 to 3 kHz
D. 3 to 6 kHz

**(B)** The more bandwidth occupied by a signal, the more frequency separation you will need from a contact currently in progress to avoid interference. CW emissions require the least bandwidth and need the least frequency separation. Most radios use narrow filters for CW reception, so you should be able to select an operating frequency within about 150 to 500 Hz from another CW station without causing interference.

**G2B05** What is the customary minimum frequency separation between SSB signals under normal conditions?

A. Between 150 and 500 Hz
B. Approximately 3 kHz
C. Approximately 6 kHz
D. Approximately 10 kHz

**(B)** The more bandwidth occupied by a signal, the more frequency separation you will need from a contact currently in progress to avoid interference. Single-sideband (SSB) signals require considerably more bandwidth than CW and therefore much more frequency separation between contacts. You will need approximately 3 kHz of separation from another contact under normal conditions to avoid causing interference.

**G2B06** What is a practical way to avoid harmful interference when selecting a frequency to call CQ on CW or phone?

A. Send "QRL?" on CW, followed by your call sign; or, if using phone, ask if the frequency is in use, followed by your call sign
B. Listen for 2 minutes before calling CQ
C. Send the letter "V" in Morse code several times and listen for a response
D. Send "QSY" on CW or if using phone, announce "the frequency is in use", then send your call and listen for a response

**(A)** After listening for a short period of time, if you do not hear another station transmitting on the frequency, it is good practice to make a short transmission asking if the frequency is in use. It may be that due to propagation you are unable to hear the transmitting station, but the listening station can hear you. On CQ, the Q signal "QRL?", and on phone, "Is the frequency in use?" followed by your call sign give the opportunity for another station to respond.

**G2B07** Which of the following complies with good amateur practice when choosing a frequency on which to initiate a call?

A. Check to see if the channel is assigned to another station
B. Identify your station by transmitting your call sign at least 3 times
C. Follow the voluntary band plan for the operating mode you intend to use
D. All of these choices are correct

**(C)** Under normal conditions, following the voluntary band plan in **Table G2B07** is a good way to choose a frequency compatible with your planned type of operating. Very crowded bands or special operating events require that you be flexible in your frequency choices.

---

### Table G2B07
### ARRL Band Plan Summary for HF Bands

| Frequency | Mode |
|---|---|
| **160 Meters (1.8 - 2.0 MHz)** | |
| 1.800- 2.000 | CW |
| 1.800- 1.810 | Digital Modes |
| 1.810 | CW QRP |
| 1.843- 2.000 | SSB, SSTV, other wideband modes |
| 1.910 | SSB QRP |
| 1.995- 2.000 | Experimental |
| 1.999- 2.000 | Beacons |
| **80 Meters (3.5-4.0 MHz):** | |
| 3.590 | RTTY/Data DX |
| 3.570-3.600 | RTTY/Data |
| 3.790-3.800 | DX window |
| 3.845 | SSTV |
| 3.885 | AM calling frequency |
| **40 Meters (7.0-7.3 MHz):** | |
| 7.040 | RTTY/Data DX |
| 7.080-7.125 | RTTY/Data |
| 7.171 | SSTV |
| 7.290 | AM calling frequency |
| **30 Meters (10.1-10.15 MHz):** | |
| 10.130-10.140 | RTTY |
| 10.140-10.150 | Packet |
| **20 Meters (14.0-14.35 MHz):** | |
| 14.070-14.095 | RTTY |
| 14.095-14.0995 | Packet |
| 14.100 | NCDXF/IARU Beacons |
| 14.1005-14.112 | Packet |
| 14.230 | SSTV |
| 14.286 | AM calling frequency |

| Frequency | Mode |
|---|---|
| **17 Meters (18.068-18.168 MHz):** | |
| 18.100-18.105 | RTTY |
| 18.105-18.110 | Packet |
| **15 Meters (21.0-21.45 MHz):** | |
| 21.070-21.110 | RTTY/Data |
| 21.340 | SSTV |
| **12 Meters (24.89-24.99 MHz):** | |
| 24.920-24.925 | RTTY |
| 24.925-24.930 | Packet |
| **10 Meters (28-29.7 MHz):** | |
| 28.000-28.070 | CW |
| 28.070-28.150 | RTTY |
| 28.150-28.190 | CW |
| 28.200-28.300 | Beacons |
| 28.300-29.300 | Phone |
| 28.680 | SSTV |
| 29.000-29.200 | AM |
| 29.300-29.510 | Satellite Downlinks |
| 29.520-29.590 | Repeater Inputs |
| 29.600 | FM Simplex |
| 29.610-29.700 | Repeater Outputs |

**G2B08**  What is the "DX window" in a voluntary band plan?

A. A portion of the band that should not be used for contacts between stations within the 48 contiguous United States
B. An FCC rule that prohibits contacts between stations within the United States and possessions on that band segment
C. An FCC rule that allows only digital contacts in that portion of the band
D. A portion of the band that has been voluntarily set aside for digital contacts only

**(A)**  Outside the United States, particularly in ITU Regions 1 and 3, amateurs share the 160 and 80 meter band with government and commercial stations. They may have very limited allocations, as well. The DX window is a section of the band where these stations may be contacted without their having to compete with stronger domestic signals. DX windows are also generally used only for contacts with stations outside the contiguous United States and Canada.

**G2B09**  Who may be the control operator of an amateur station transmitting in RACES to assist relief operations during a disaster?

A. Only a person holding an FCC issued amateur operator license
B. Only a RACES net control operator
C. A person holding an FCC issued amateur operator license or an appropriate government official
D. Any control operator when normal communication systems are operational

**(A)**  The control operator of a RACES station must have an FCC-issued amateur operator license and be certified by a civil defense organization as a member. [97.407(a)]

**G2B10**  When may the FCC restrict normal frequency operations of amateur stations participating in RACES?

A. When they declare a temporary state of communication emergency
B. When they seize your equipment for use in disaster communications
C. Only when all amateur stations are instructed to stop transmitting
D. When the President's War Emergency Powers have been invoked

**(D)**  If the War Emergency Powers have been activated, RACES stations are restricted to operations in the frequency ranges listed in §97.407(b).

**G2B11** **What frequency should be used to send a distress call?**

A. Whatever frequency has the best chance of communicating the distress message
B. Only frequencies authorized for RACES or ARES stations
C. Only frequencies that are within your operating privileges
D. Only frequencies used by police, fire or emergency medical services

**(A)** When normal communications are not available and the immediate safety of human life or protection of property is involved, all of the normal rules for an amateur station are suspended so that you can obtain assistance. This means that any method of communication, on any frequency, and at any power output, may be used to communicate and resolve the emergency. It doesn't matter if the distress is personal to the station or a general disaster. Just be sure you have a real emergency! [97.405]

**G2B12** **When is an amateur station allowed to use any means at its disposal to assist another station in distress?**

A. Only when transmitting in RACES
B. At any time when transmitting in an organized net
C. At any time during an actual emergency
D. Only on authorized HF frequencies

**(C)** No FCC rule prevents an amateur station from using any means of radiocommunications at its disposal to assist a station in distress. [97.405(b)]

## G2C  CW operating procedures and procedural signals, Q signals and common abbreviations; full break in

**G2C01** **Which of the following describes full break-in telegraphy (QSK)?**

A. Breaking stations send the Morse code prosign BK
B. Automatic keyers are used to send Morse code instead of hand keys
C. An operator must activate a manual send/receive switch before and after every transmission
D. Transmitting stations can receive between code characters and elements

**(D)** Full break-in telegraphy allows you to receive signals between your transmitted Morse code dots and dashes and between words. The advantage is that if you are sending a long message, the receiving station can send back to you (break in) and stop you for repeats of missed words. QSK is the Q signal used to describe this type of operation.

**G2C02** **What should you do if a CW station sends "QRS"?**

A. Send slower
B. Change frequency
C. Increase your power
D. Repeat everything twice

**(A)** QRS is the Q signal that means "Send slower". To ask if you should send slower, send QRS? Conversely, to increase speed, QRQ is used.

**G2C03** **What does it mean when a CW operator sends "KN" at the end of a transmission?**

A. Listening for novice stations
B. Operating full break-in
C. Listening only for a specific station or stations
D. Closing station now

**(C)** KN is an example of a CW prosign, procedural signals that help coordinate the exchange of messages and the beginning and ending of transmissions. The patterns of dots and dashes that make up prosigns are described by a pair of regular letters that, if sent together without a pause, are equivalent to the prosign. (Prosigns are often written with a line over the letters to indicate they are sent with no spaces between them as a single character.)

**G2C04** **What does it mean when a CW operator sends "CL" at the end of a transmission?**

A. Keep frequency clear
B. Operating full break-in
C. Listening only for a specific station or stations
D. Closing station

**(D)** CL is an example of a CW prosign, procedural signals that help coordinate the exchange of messages and the beginning and ending of transmissions. The patterns of dots and dashes that make up prosigns are described by a pair of regular letters that, if sent together without a pause, are equivalent to the prosign. (Prosigns are often written with a line over the letters to indicate they are sent with no spaces between them as a single character.)

**G2C05** **What is the best speed to use answering a CQ in Morse Code?**

A. The fastest speed at which you are comfortable copying
B. The speed at which the CQ was sent
C. A slow speed until contact is established
D. 5 wpm, as all operators licensed to operate CW can copy this speed

**(B)** An operator calling CQ is assumed to be sending at a speed at which he or she feels comfortable receiving. Responding at a significantly higher speed is impolite and may be embarrassing to the other operator if they are unable to copy your response. If you are uncomfortable responding at the sending station's speed, send at the highest rate at which you are comfortable receiving. It is good practice to respond to calling stations at their sending speed, if it is significantly slower.

**G2C06**  What does the term "zero beat" mean in CW operation?

A. Matching the speed of the transmitting station
B. Operating split to avoid interference on frequency
C. Sending without error
D. Matching your transmit frequency to the frequency of a received signal.

**(D)** Zero beat means to match the frequency of the transmitting station. When separate receivers and transmitters were the norm, a transmitter's frequency had to be adjusted to match the received signal's frequency. This was done by spotting — turning on the transmitter's low power stages and listening for that signal in the receiver. When the beat frequency between the desired signal and the transmitter's spotting signal reached zero frequency or zero beat, the transmitter signal and the received signal were on matching frequencies.

**G2C07**  When sending CW, what does a "C" mean when added to the RST report?

A. Chirpy or unstable signal
B. Report was read from S meter reading rather than estimated
C. 100 percent copy
D. Key clicks

**(A)** An RST with "C" appended, such as 579C, indicates that the signal is being received with chirp, a short frequency shift as the transmitter stabilizes after keying. It's a very distinctive sound and is caused by the transmitter's oscillator changing frequency when the key is closed. This can be due to oscillator circuit design or poor regulation of the transmitter power supply.

**G2C08**  What prosign is sent to indicate the end of a formal message when using CW?

A. SK
B. BK
C. AR
D. KN

**(C)** The ARRL National Traffic System has established specific procedures for passing formal written messages by Amateur Radio. Even if you don't participate in traffic nets, it is a good idea to be familiar with the procedures for handling such messages. It can be especially helpful in an emergency for a number of reasons. By following the standard procedures it is more likely that an emergency message will be transmitted (and received) correctly. When sending formal messages using Morse code (CW), you send the message preamble, the address, message body and signature. To indicate that this is the end of the message, send the CW procedural signal (prosign) "AR" to show clearly that all the information has been sent. When the receiving station has accurately recorded the entire message, they will acknowledge receipt of the message by sending "QSL" or simply "R" for "received." (see also the discussion for G2F03)

**G2C09**  What does the Q signal "QSL" mean?

A. Send slower
B. We have already confirmed by card
C. I acknowledge receipt
D. We have worked before

**(C)** QSL is the Q signal that means "I acknowledge receipt". Informally, it is often used to indicate that a transmission was received and understood. QSL cards are exchanged to confirm that a contact was made.

**G2C10**  What does the Q signal "QRQ" mean?

A. Slow down
B. Send faster
C. Zero beat my signal
D. Quitting operation

**(B)** QRQ is the Q signal that means "Send faster." To ask if you should send faster, send QRQ? Conversely, to decrease speed, QRS is used.

**G2C11**  What does the Q signal "QRV" mean?

A. You are sending too fast
B. There is interference on the frequency
C. I am quitting for the day
D. I am ready to receive messages

**(D)** QRV is the Q signal that means "I am ready to copy" and indicates that the station with the message may begin transmitting. QRV is used whether the message is formal traffic or just regular conversation.

## G2D    Amateur Auxiliary; minimizing interference; HF operations

**G2D01**  What is the Amateur Auxiliary to the FCC?

A. Amateur volunteers who are formally enlisted to monitor the airwaves for rules violations
B. Amateur volunteers who conduct amateur licensing examinations
C. Amateur volunteers who conduct frequency coordination for amateur VHF repeaters
D. Amateur volunteers who use their station equipment to help civil defense organizations in times of emergency

**(A)** The purpose of the Amateur Auxiliary is to help ensure amateur self-regulation and see that amateurs follow the FCC rules properly. The Amateur Auxiliary volunteers deal only with amateur-to-amateur interference and improper operation. The other answer choices describe other Amateur Radio activities. Amateur volunteers who conduct licensing examinations are called Volunteer Examiners (VEs). Amateurs in charge of frequency coordination for repeaters are called Frequency Coordinators. Amateurs who help civil defense organizations in times of emergency are members of the Radio Amateur Civil Emergency Service (RACES).

**G2D02**  Which of the following are the objectives of the Amateur Auxiliary?

A. To conduct efficient and orderly amateur licensing examinations
B. To encourage amateur self regulation and compliance with the rules
C. To coordinate repeaters for efficient and orderly spectrum usage
D. To provide emergency and public safety communications

**(B)** Many amateurs also volunteer to help provide emergency and public safety communications as members of ARRL's Amateur Radio Emergency Service (ARES). (see the discussion for question G2D01)

**G2D03**  What skills learned during "hidden transmitter hunts" are of help to the Amateur Auxiliary?

A. Identification of out of band operation
B. Direction finding used to locate stations violating FCC Rules
C. Identification of different call signs
D. Hunters have an opportunity to transmit on non-amateur frequencies

**(B)** Friendly competitions to locate hidden transmitters, sometimes called "fox hunts" or "bunny hunts", allow participants to practice their radio direction-finding skills which are useful in locating harmful interference sources. The Amateur Auxiliary can use "Fox Hunters" to document interference cases and report them to the proper enforcement bureau. Fox hunts also make everyone aware that there is a plan in place to find and eliminate an interference source.

**G2D04** Which of the following describes an azimuthal projection map?

A. A world map that shows accurate land masses
B. A world map projection centered on a particular location
C. A world map that shows the angle at which an amateur satellite crosses the equator
D. A world map that shows the number of degrees longitude that an amateur satellite appears to move westward at the equator with each orbit

**(B)** An azimuthal map, or azimuthal-equidistant projection map, is also called a great circle map. When this type of map is centered on your location, a straight line is equivalent to stretching a string between two points on a globe, and will give you the shortest distance between two points. This type of map is used for determining the direction to point your antenna for short-path communications. (A compass bearing 180 degrees different from the short-path direction will indicate the direction to point your antenna for long-path communications.)

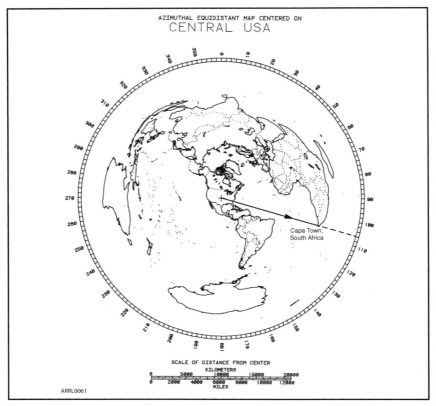

**Figure G2-1 — This azimuthal equidistant world map centered on the central United States shows how an amateur at that location would point a beam antenna for a contact with a station in Cape Town, South Africa.**

**G2D05** When is it permissible to communicate with amateur stations in countries outside the areas administered by the Federal Communications Commission?

A. Only when the foreign country has a formal third party agreement filed with the FCC
B. When the contact is with amateurs in any country except those whose administrations have notified the ITU that they object to such communications
C. When the contact is with amateurs in any country as long as the communication is conducted in English
D. Only when the foreign country is a member of the International Amateur Radio Union

**(B)** U.S. amateurs are permitted to contact amateurs in any other country. There have been a very few instances in which a government has prohibited contact between its amateurs and those of a particular country. [97.111(a)(1)]

**G2D06** How is a directional antenna pointed when making a "long-path" contact with another station?

A. Toward the rising Sun
B. Along the gray line
C. 180 degrees from its short-path heading
D. Toward the north

**(C)** The shortest direct route, or great-circle path between two points, is called the short-path. If a directional antenna is pointed in exactly the opposite direction, 180 degrees different from the short-path direction, communications can be attempted on the long-path. Long-path communication may be available when the more direct short path is closed. Because of the higher number of hops required, long path often works best when the path is across the ocean, a good reflector of HF signals.

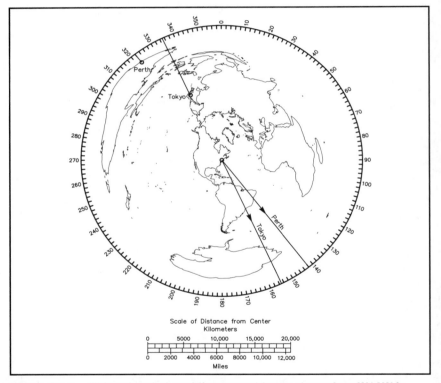

**Figure G2-2 — This azimuthal equidistant world map shows how W1AW in Newington, Connecticut would point a beam antenna to make a long-path contact with Tokyo, Japan or Perth, Australia. Notice that the paths in both cases lie almost entirely over water, rather than over land masses.**

**G2D07** Which of the following is required by the FCC rules when operating in the 60 meter band?

A. If you are using other than a dipole antenna, you must keep a record of the gain of your antenna
B. You must keep a log of the date, time, frequency, power level and stations worked
C. You must keep a log of all third party traffic
D. You must keep a log of the manufacturer of your equipment and the antenna used

**(A)** The FCC Rules for operating on the amateur 60 meter band are quite different from the rules for any other amateur band. One significant difference is the requirement to transmit with no more than 50 watts effective radiated power (ERP). ERP is a measurement of power as compared to that radiated from a dipole. If you are using a half-wavelength dipole antenna, you can use up to 50 W PEP from your transmitter on that band. If you are using an antenna that has some gain compared to a dipole, then you will have to reduce your transmitter power accordingly. If the antenna has a gain of 3 dBd (3 dB compared to a dipole) then you would have to reduce transmitter power by half (to 25 W PEP). The FCC requires you to keep a record of your antenna gain, if it is more than a dipole. This record can either be from the manufacturer's data, from calculations of the gain or from gain measurements. [97.303(s)]

**G2D08** Why do many amateurs keep a log even though the FCC doesn't require it?

A. The ITU requires a log of all international contacts
B. The ITU requires a log of all international third party traffic
C. The log provides evidence of operation needed to renew a license without retest
D. To help with a reply if the FCC requests information

**(D)** While useful, the FCC does not require you to keep a record (log) of your transmissions. It can be fun to keep a log, though, and then look back years later at the contacts you made. A log can also help document when your station was on the air and who was the control operator. You must give permission before a visiting amateur may operate your station. If you designate another amateur to be the control operator of your station, you both share the responsibility for the proper operation of the station. Unless your station records (log) show otherwise, the FCC will assume you were the control operator any time your station was operated. [97.103(b)]

**G2D09**   What information is traditionally contained in a station log?

A.  Date and time of contact
B.  Band and/or frequency of the contact
C.  Call sign of station contacted and the signal report given
D.  All of these choices are correct

**(D)** You can keep any information in your log that you would like to remember later. At a minimum, most amateurs keep a record of the date and time of each contact as well as the frequency or band of the contact. The station call sign, mode and the RST signal report given and received are also normally recorded. Many amateurs also record the name of the other operator as well as his or her location.

**G2D10**   What is QRP operation?

A.  Remote piloted model control
B.  Low power transmit operation
C.  Transmission using Quick Response Protocol
D.  Traffic relay procedure net operation

**(B)** QRP is a Q signal that means "lower your transmitter power." Many amateurs enjoy using low power levels for the challenge, the relative simplicity of the equipment, and sometimes to reduce interference. The generally accepted level for "QRP" power is 5 watts of transmitter output on CW and 10 W PEP output on phone.

**G2D11**   Which HF antenna would be the best to use for minimizing interference?

A.  A quarter wave vertical antenna
B.  An isotropic antenna
C.  A unidirectional antenna
D.  An omnidirectional antenna

**(C)** Most of the time we think about using a beam antenna to send more of our transmitted power toward the desired station. A beam antenna can also be an effective tool for fighting received interference. For example, you might be having a conversation with another amateur but a strong signal from another station on a nearby frequency is causing a bit of interference. With a beam antenna you might be able to turn the antenna so the interfering station is off the side or back of your antenna. In that case the antenna will not receive as strong a signal from the interfering station. Ideally, you would have an antenna that sends and receives signals in only one direction, rejecting signals in all other directions. This is called a unidirectional antenna. Most Yagi antennas send and receive some signal off the back and sides of the antenna but it is significantly less than the amount that is sent and received from the front.

## G2E  Digital operating: procedures, procedural signals and common abbreviations

**G2E01**  Which mode is normally used when sending an RTTY signal via AFSK with an SSB transmitter?

A. USB
B. DSB
C. CW
D. LSB

**(D)** Lower sideband (LSB) is used by convention for RTTY signals on all bands. There is no technical reason why LSB is preferred over USB for RTTY signals.

**G2E02**  How many data bits are sent in a single PSK31 character?

A. The number varies
B. 5
C. 7
D. 8

**(A)** One of the interesting properties of the PSK31 radioteletype mode is that it uses a character code called Varicode. Developed by Peter Martinez, G3PLX, Varicode uses shorter character lengths for the more common characters and longer codes for less common characters. (Morse code also uses variable length characters.) Thus, the number of data bits per character depends on which character is being sent. This is quite different from other digital modes that use fixed-length codes. For example, all RTTY Baudot code characters have five data bits.

**G2E03**  What part of a data packet contains the routing and handling information?

A. Directory
B. Preamble
C. Header
D. Footer

**(C)** Packet radio gets its name from the concept of breaking the information to be transmitted into small pieces, called packets. Each packet can be transmitted and make its way through the network independently. The entire message is then reassembled at the receiving station. Each packet consists of a header that contains several sets (fields) of information about the packet. The address field of the header contains information about the station for which the message is intended, the sending station and possibly specific relay stations. A control field contains information about the type of data being sent. A frame number allows the receiving station reassemble the entire message in order. The data field of the packet contains the actual data being sent. After the data field there is a frame check sequence (FCS) or cyclic redundancy check (CRC) field that is used for error detection.

**G2E04**   What segment of the 20 meter band is most often used for data transmissions?

A. 14.000 - 14.050 MHz
B. 14.070 - 14.100 MHz
C. 14.150 - 14.225 MHz
D. 14.275 - 14.350 MHz

**(B)** The FCC's rules specify where RTTY and data transmissions allowed, but the band plans tell you where such signals are usually found. The band plan in **Table G2E04** calls for RTTY and data operation from 14.070 to 14.100 MHz on 20 meters.

---

**Table G2E04**
**ARRL Band Plan for RTTY/Data Frequencies**

| Band | RTTY/Data (Meters) | Frequencies (MHz) |
|------|------|------|
| 160 | 1.800 - | 1.810 |
| 80 | 3.570 - | 3.600 |
| 40 | 7.080 - | 7.125 |
| 30 | 10.130 - | 10.140 |
| 20 | 14.070 - | 14.0995 |
| 17 | 18.100 - | 18.105 |
| 15 | 21.070 - | 21.110 |
| 12 | 24.920 - | 24.925 |
| 10 | 28.070 - | 28.150 |

---

**G2E05**   Which of the following describes Baudot code?

A. A 7-bit code, with start, stop and parity bits
B. A code using error detection and correction
C. A 5-bit code, with additional start and stop bits
D. A code using SELCAL and LISTEN

**(C)** The Baudot code used for radioteletype (RTTY) has five data bits per character. This means there are only 32 possible character combinations. The code only transmits upper case letters. "LTRS" and "FIGS" characters select the alphabet table or a numbers and punctuation table to provide additional characters. RTTY is an asynchronous communications mode, so each character must also include a start and a stop bit.

**G2E06** **What is the most common frequency shift for RTTY emissions in the amateur HF bands?**

A. 85 Hz
B. 170 Hz
C. 425 Hz
D. 850 Hz

**(B)** RTTY operation on the HF bands uses frequency-shift keying (FSK) to convey the information. The transmitted signal shifts between two frequencies, called the MARK and SPACE frequencies. The two frequencies used are normally 170 Hz apart for HF communications.

**G2E07** **What does the abbreviation "RTTY" stand for?**

A. Returning to you
B. Radioteletype
C. A general call to all digital stations
D. Repeater transmission type

**(B)** RTTY stands for radioteletype. The FCC's name for RTTY emissions is narrow-band, direct printing telegraphy.

**G2E08** **What segment of the 80 meter band is most commonly used for data transmissions?**

A. 3570 – 3600 kHz
B. 3500 – 3525 kHz
C. 3700 – 3750 kHz
D. 3775 – 3825 kHz

**(A)** The FCC's rules specify where RTTY and data transmissions allowed, but the band plans tell you where such signals are usually found. The band plan calls for RTTY operation from 3570 to 3600 kHz on 80 meters. See the band plans in the explanation for question G2E04.

**G2E09** **In what segment of the 20 meter band are most PSK31 operations commonly found?**

A. At the bottom of the slow-scan TV segment, near 14.230 MHz
B. At the top of the SSB phone segment near 14.325 MHz
C. In the middle of the CW segment, near 14.100 MHz
D. Below the RTTY segment, near 14.070 MHz

**(D)** PSK signals are generally found in the vicinity of 14.070 MHz on the 20 meter band at the bottom of the RTTY area listed in **Table G2E04**.

**G2E10**   Question has been withdrawn.

**G2E11**   What does the abbreviation "MFSK" stand for?
   A. Manual Frequency Shift Keying
   B. Multi (or Multiple) Frequency Shift Keying
   C. Manual Frequency Sideband Keying
   D. Multi (or Multiple) Frequency Sideband Keying

**(B)**   MFSK stands for Multi-Frequency Shift Keying. See question G2E10.

**G2E12**   How does the receiving station respond to an ARQ data mode packet containing errors?
   A. Terminates the contact
   B. Requests the packet be retransmitted
   C. Sends the packet back to the transmitting station
   D. Requests a change in transmitting protocol

**(B)**   ARQ stands for Automatic Repeat Request. In an ARQ digital mode, if a transmitted packet is received with errors, the receiving station may request a re-transmission of the packet automatically.

**G2E13**   In the PACTOR protocol, what is meant by an NAK response to a transmitted packet?
   A. The receiver is requesting the packet be re-transmitted
   B. The receiver is reporting the packet was received without error
   C. The receiver is busy decoding the packet
   D. The entire file has been received correctly

**(A)**   PACTOR is an ARQ (Automatic Repeat Request) digital mode in which the receiving station sends a response to the transmitting station indicating whether the data was received correctly or not. An ACK (Acknowledge) response means the data was received correctly. The NAK (Not Acknowledge) response means that errors were detected in the received data.

# Radio Wave Propagation

Your General class exam (Element 3) will consist of 35 questions taken from the General class question pool as prepared by the Volunteer Examiner Coordinators' Questions Pool Committee. A certain number of questions are taken from each of the 10 subelements. There will be 3 questions from the subelement shown in this chapter. These questions are divided into 3 groups, labeled G3A through G3C.

## SUBELEMENT G3 — RADIO WAVE PROPAGATION
### [3 Exam Questions — 3 Groups]

**G3A** **Sunspots and solar radiation; ionospheric disturbances; propagation forecasting and indices**

**G3A01** What is the sunspot number?
A. A measure of solar activity based on counting sunspots and sunspot groups
B. A 3 digit identifier which is used to track individual sunspots
C. A measure of the radio flux from the Sun measured at 10.7 cm
D. A measure of the sunspot count based on radio flux measurements

**(A)** A number of observatories around the world measure solar activity. A weighted average of this data is used to determine the International Sunspot Number (ISN) for each day. These daily sunspot counts are used to produce monthly and yearly average values. The average values are used to see trends and patterns in the measurements.

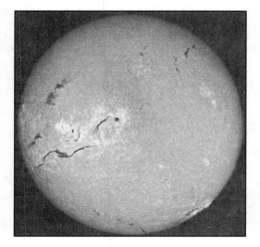

**Figure G3-1 — Much more than sunspots can be seen when the Sun is viewed through selective optical filters. This photo was taken through a hydrogen-alpha filter, which passes a narrow band of light wavelengths at 6562 angstroms (1 angstrom = 1 × 10⁻¹⁰ meters). The bright patches are active areas around and often between sunspots. Dark irregular lines are filaments of activity having no central core. Faint magnetic field lines are visible around a large central sunspot group near the disc center. (Photo courtesy of Sacramento Peak Observatory, Sunspot, New Mexico.)**

**G3A02** What effect does a Sudden Ionospheric Disturbance have on the daytime ionospheric propagation of HF radio waves?

A. It enhances propagation on all HF frequencies
B. It disrupts signals on lower frequencies more than those on higher frequencies
C. It disrupts communications via satellite more than direct communications
D. None, because only areas on the night side of the Earth are affected

**(B)** A sudden ionospheric disturbance (SID) is often the result of solar flares that release large amounts of radiation. Ultraviolet and X-ray radiation from the Sun travels at the speed of light, reaching the Earth in about eight minutes. When this radiation reaches the Earth, the level of ionization in the ionosphere increases rapidly. This causes D-layer absorption of radio waves to increase significantly. Absorption of radio signals in the D layer is always stronger at lower frequencies, affecting lower frequency signals more than higher frequency signals.

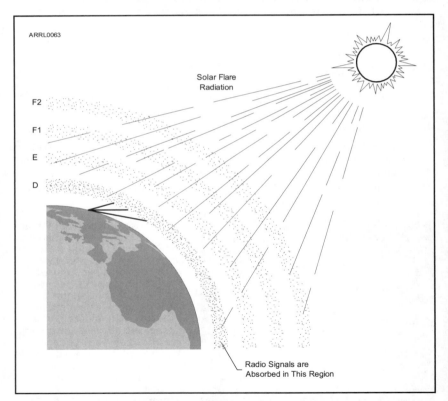

Figure G3-2 — Approximately eight minutes after a solar flare occurs on the Sun, the ultraviolet and X-ray radiation released by the flare reaches the Earth. This radiation causes increased ionization and radio-wave absorption in the D region of the ionosphere.

**G3A03** Approximately how long does it take the increased ultraviolet and X-ray radiation from solar flares to affect radio-wave propagation on the Earth?

A. 28 days
B. 1 to 2 hours
C. 8 minutes
D. 20 to 40 hours

**(C)** Ultraviolet and X-ray radiation from the Sun travels at the speed of light, reaching the Earth in about eight minutes.

**G3A04** Which of the following amateur radio HF frequencies are least reliable for long distance communications during periods of low solar activity?

A. 3.5 MHz and lower
B. 7 MHz
C. 10 MHz
D. 21 MHz and higher

**(D)** The higher the frequency, the more ionization is needed in the ionosphere in order to refract (bend) the radio signal back to the Earth. When solar activity is low, the higher frequencies will pass through the ionosphere into space instead of being refracted back to Earth. During periods of low solar activity, the 15 meter (21 MHz), 12 meter (24.9 MHz) and 10 meter (28 MHz) bands are the least reliable HF bands for long distance communication.

**G3A05** What is the solar-flux index?

A. A measure of the highest frequency that is useful for ionospheric propagation between two points on the Earth
B. A count of sunspots which is adjusted for solar emissions
C. Another name for the American sunspot number
D. A measure of solar radiation at 10.7 cm

**(D)** Solar flux is the radio energy coming from the Sun. High levels of solar energy produce greater ionization in the ionosphere. The solar flux measurement is taken daily by measuring radio energy from the Sun at 2800 MHz which is a wavelength of 10.7 cm. The measurement is then converted into the solar flux index. Higher values of the solar flux index correspond to higher values of solar flux. The solar-flux measurement may be taken under any weather conditions — the Sun does not have to be visible, as for determining the sunspot number. The radio energy measurement is converted to an open-ended numeric index with a minimum value of 65 (for the minimum amount of energy). Higher values of the solar flux index indicate higher levels of solar activity.

**G3A06**  What is a geomagnetic storm?

A. A sudden drop in the solar-flux index
B. A thunderstorm which affects radio propagation
C. Ripples in the ionosphere
D. A temporary disturbance in the Earth's magnetosphere

**(D)** Geomagnetic disturbances result when charged particles from a solar flare reach the Earth. When these charged particles reach the Earth's magnetic field, they are deflected toward the North and South poles. Radio communications along higher-latitude paths (latitudes greater than about 45 degrees) will be more affected than paths closer to the equator. The charged particles from the Sun may make the F-region seem to disappear or seem to split into many layers, degrading or completely blacking out long-distance radio communications.

**G3A07**  At what point in the solar cycle does the 20 meter band usually support worldwide propagation during daylight hours?

A. At the summer solstice
B. Only at the maximum point of the solar cycle
C. Only at the minimum point of the solar cycle
D. At any point in the solar cycle

**(D)** Even at the minimum point of the solar cycle, world-wide propagation is usually possible on the 20 meter band. As solar activity increases, the band will remain open for longer periods and with stronger signal strengths. For this reason, 20 meters is a favorite band for "DXers".

**G3A08**  Which of the following effects can a geomagnetic storm have on radio-wave propagation?

A. Improved high-latitude HF propagation
B. Degraded high-latitude HF propagation
C. Improved ground-wave propagation
D. Improved chances of UHF ducting

**(B)** Geomagnetic disturbances result when charged particles from a solar flare reach the Earth. When these charged particles reach the Earth's magnetic field, they are deflected toward the North and South poles. Radio communications along higher-latitude paths (latitudes greater than about 45 degrees) will be more affected than paths closer to the equator. The charged particles from the Sun may make the F-region seem to disappear or seem to split into many layers, degrading or completely blacking out long-distance radio communications.

**G3A09** What effect do high sunspot numbers have on radio communications?

A. High-frequency radio signals become weak and distorted
B. Frequencies above 300 MHz become usable for long-distance communication
C. Long-distance communication in the upper HF and lower VHF range is enhanced
D. Microwave communications become unstable

**(C)** When sunspot numbers are high, there is a significant amount of solar activity and there will be more ionization of the ionosphere. The more the ionosphere is ionized, the higher the frequency of radio signals that may be used for long-distance communication. During the peak of a sunspot cycle, the 20 meter (14 MHz) band will be open around the world even through the night. During an unusually good sunspot cycle, even the 6 meter (50 MHz) band can become usable for long-distance communication.

**G3A10** What causes HF propagation conditions to vary periodically in a 28-day cycle?

A. Long term oscillations in the upper atmosphere
B. Cyclic variation in the Earth's radiation belts
C. The Sun's rotation on its axis
D. The position of the Moon in its orbit

**(C)** It takes approximately 28 days for the Sun to rotate on its axis. Since active areas on the Sun may persist for more than one rotation, you can expect good propagation conditions to recur approximately every 28 days.

**G3A11** Approximately how long is the typical sunspot cycle?

A. 8 minutes
B. 40 hours
C. 28 days
D. 11 years

**(D)** The internal dynamics of the Sun cause its activity to vary in a cycle lasting approximately 11 years. Sunspots are one indication of solar activity and since they were the earliest phenomenon observed on the Sun, the cycle is called the sunspot cycle.

**G3A12**  What does the K-index indicate?

A.  The relative position of sunspots on the surface of the Sun
B.  The short term stability of the Earth's magnetic field
C.  The stability of the Sun's magnetic field
D.  The solar radio flux at Boulder, Colorado

**(B)**  The K-index represents readings of the Earth's geomagnetic field, updated every three hours at Boulder, Colorado. K-index values indicate the stability of the Earth's geomagnetic field. Steady values indicate a stable geomagnetic field, while rising values indicate an active geomagnetic field. The K-index trends are important indicators of changing propagation conditions. Rising K-index values are generally bad news for HF propagation, especially for propagation paths involving latitudes above 30° north. Values of 4 and rising warn of conditions associated with auroras and degraded HF propagation.

**G3A13**  What does the A-index indicate?

A.  The relative position of sunspots on the surface of the Sun
B.  The amount of polarization of the Sun's electric field
C.  The long term stability of the Earth's geomagnetic field
D.  The solar radio flux at Boulder, Colorado

**(C)**  The A-index is a daily figure for the state of activity of the Earth's magnetic field. The A-index tells you mainly about yesterday's conditions, but it is very revealing when charted regularly, because geomagnetic disturbances nearly always recur at four-week intervals. (It takes the Sun 28 days to rotate once on its axis.)

**G3A14**  How are radio communications usually affected by the charged particles that reach the Earth from solar coronal holes?

A.  HF communications are improved
B.  HF communications are disturbed
C.  VHF/UHF ducting is improved
D.  VHF/UHF ducting is disturbed

**(B)**  The corona is the Sun's outer layer. Temperatures in the corona are typically about two million degrees Celsius, but can be more than four million degrees Celsius above an active sunspot region. A coronal hole is an area of somewhat lower temperature. Matter ejected through such a "hole" is in the form of plasma, a highly ionized gas made up of electrons, protons and neutral particles. The plasma travels at speeds up to two million miles per hour, and if the "jet" of material is directed toward the Earth it can result in a geomagnetic storm on Earth, disrupting HF communications.

**G3A15** How long does it take charged particles from coronal mass ejections to affect radio-wave propagation on the Earth?

A. 28 days
B. 14 days
C. 4 to 8 minutes
D. 20 to 40 hours

**(D)** The corona is the Sun's outer layer. Temperatures in the corona are typically about two million degrees Celsius, but can be more than four million degrees Celsius above an active sunspot region. A coronal hole is an area of somewhat lower temperature. Matter ejected through such a "hole" is in the form of plasma, a highly ionized gas made up of electrons, protons and neutral particles. The plasma travels at speeds of two million miles per hour or more, so it can take about 20 to 40 hours for the plasma to travel the 93 million miles to Earth.

**G3A16** What is a possible benefit to radio communications resulting from periods of high geomagnetic activity?

A. Aurora that can reflect VHF signals
B. Higher signal strength for HF signals passing through the polar regions
C. Improved HF long path propagation
D. Reduced long delayed echoes

**(A)** When the plasma, or charged particles, from a coronal mass ejection reaches the Earth it interacts with the Earth's magnetic field. The charged particles follow magnetic field lines into the Earth's atmosphere near the North and South magnetic poles, producing visible aurora borealis at northern latitudes and aurora australis in the south. VHF operators look forward to such conditions because radio signals can be reflected from auroral "patches," making long-distance contacts possible.

## G3B  Maximum Usable Frequency; Lowest Usable Frequency; Propagation

**G3B01**  How might a sky-wave signal sound if it arrives at your receiver by both short path and long path propagation?

A. Periodic fading approximately every 10 seconds
B. Signal strength increased by 3 dB
C. The signal might be cancelled causing severe attenuation
D. A well-defined echo might be heard

**(D)** Normally, you will expect radio signals to arrive at your station by following the shortest possible path between you and the transmitting station. This is called short-path propagation. Signals that might have arrived from the opposite direction, 180 degrees different from the short-path signals are normally so weak that you would probably not hear them. Signals that arrive 180 degrees from the short path are called long-path signals. When propagation conditions are suitable, the long-path signals may be strong enough to support communication. In fact, there are times when the long-path propagation may be even better than the short-path propagation. Stations with directional antennas can point their antennas directly away from each other to communicate. (This is not simply communicating using signal radiated "off the back" of the antennas.) If you are listening to signals on your receiver and you hear a well-defined echo, even if it is a weak echo, the chances are you are hearing signals arrive at your station over the long path. The slightly longer time it takes the signals to travel the longer distance around the Earth results in a slight delay when compared to the direct, short-path signals. This is a good indication that you may be able to point your antenna directly away from the received station to communicate.

**G3B02**  Which of the following is a good indicator of the possibility of sky-wave propagation on the 6 meter band?

A. Short skip sky-wave propagation on the 10 meter band
B. Long skip sky-wave propagation on the 10 meter band
C. Severe attenuation of signals on the 10 meter band
D. Long delayed echoes on the 10 meter band

**(A)** As the maximum usable frequency (MUF) for a given path increases, the ionosphere also supports shorter single-hop distance at lower frequencies. Suppose you are operating on the 10-meter band, and are contacting stations that are 800 to 1000 miles away. After making a few more contacts you start to notice that you are contacting stations only about 500 miles away, and then you notice that you are contacting stations even closer, perhaps only out to a few hundred miles. This can be an excellent indication that the MUF for the longer-path stations has moved up to a higher frequency, perhaps even above 50 MHz. It is a good time to check for a band opening on 6 meters!

**G3B03**   Which of the following applies when selecting a frequency for lowest attenuation when transmitting on HF?

A. Select a frequency just below the MUF
B. Select a frequency just above the LUF
C. Select a frequency just below the critical frequency
D. Select a frequency just above the critical frequency

**(A)** Ionospheric absorption (attenuation) is lowest just below the Maximum Usable Frequency (MUF). Use a frequency just below the MUF for the highest received signal strength. (see also the discussion for question G3B05)

**G3B04**   What is a reliable way to determine if the Maximum Usable Frequency (MUF) is high enough to support skip propagation between your station and a distant location on frequencies between 14 and 30 MHz?

A. Listen for signals from an international beacon
B. Send a series of dots on the band and listen for echoes from your signal
C. Check the strength of TV signals from Western Europe
D. Check the strength of signals in the MF AM broadcast band

**(A)** Beacon stations transmit signals so that amateur operators can evaluate propagation conditions. By listening for beacon stations from Western Europe, you will be able to determine if the MUF is high enough for 10 meter communications to that area. (see also the discussion for question G3B05)

**G3B05** What usually happens to radio waves with frequencies below the Maximum Usable Frequency (MUF) and above the Lowest Usable Frequency (LUF) when they are sent into the ionosphere?

A. They are bent back to the Earth
B. They pass through the ionosphere
C. They are amplified by interaction with the ionosphere
D. They are bent and trapped in the ionosphere to circle the Earth

**(A)** The Maximum Usable Frequency (MUF) relates to a particular desired destination. The MUF is the highest frequency that will allow the radio wave to reach its desired destination using E or F-region propagation. There is no single MUF for a given transmitter location; it will vary depending on the direction and distance to the station you are attempting to contact. Frequencies lower than the MUF are generally bent back to Earth. Frequencies higher than the MUF will pass through the ionosphere instead of being bent back to the Earth.

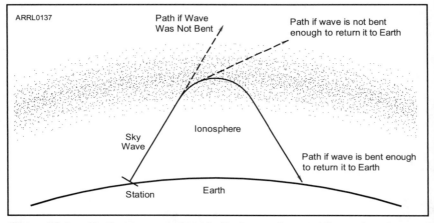

Figure G3-3 — Radio waves are bent in the ionosphere, so they return to Earth far from the transmitter. If the radio wave is not bent (refracted) enough in the ionosphere, it will pass into space rather than returning to Earth.

**G3B06** What usually happens to radio waves with frequencies below the Lowest Usable Frequency (LUF)?

A. They are bent back to the Earth
B. They pass through the ionosphere
C. They are completely absorbed by the ionosphere
D. They are bent and trapped in the ionosphere to circle the Earth

**(C)** The Lowest Usable Frequency (LUF) is the frequency below which ionospheric absorption attenuates the radio signals to below the atmospheric noise levels. Since absorption increases with decreasing frequency, signals at frequencies below the LUF can not be received via sky-wave communication.

## G3B07  What does LUF stand for?

A. The Lowest Usable Frequency for communications between two points
B. The Longest Universal Function for communications between two points
C. The Lowest Usable Frequency during a 24 hour period
D. The Longest Universal Function during a 24 hour period

(A) The Lowest Usable Frequency (LUF) is the frequency below which ionospheric absorption attenuates the radio signals to below the atmospheric noise levels. Since absorption increases with decreasing frequency, signals at frequencies below the LUF can not be received via sky-wave communication.

## G3B08  What does MUF stand for?

A. The Minimum Usable Frequency for communications between two points
B. The Maximum Usable Frequency for communications between two points
C. The Minimum Usable Frequency during a 24 hour period
D. The Maximum Usable Frequency during a 24 hour period

(B) The Maximum Usable Frequency (MUF) relates to a particular desired destination. The MUF is the highest frequency that will allow the radio wave to reach its desired destination using E or F-region propagation. There is no single MUF for a given transmitter location; it will vary depending on the direction and distance to the station you are attempting to contact. Frequencies lower than the MUF are generally bent back to Earth. Frequencies higher than the MUF will pass through the ionosphere instead of being bent back to the Earth.

**G3B09**   What is the approximate maximum distance along the Earth's surface that is normally covered in one hop using the F2 region?

A.  180 miles
B.  1,200 miles
C.  2,500 miles
D.  12,000 miles

**(C)**  Layers in the F region form and decay in correlation with the daily passage of the sun. The F1 and F2 layers form when the F region splits into two parts due to high radiation from the Sun, recombining into a single F layer at night. The more solar radiation the F region receives, the more it is ionized so it reaches maximum ionization shortly after noon during the summertime. The ionization tapers off very gradually towards sunset and the F2 layer remains usable into the night. The F2 region is the highest of the ionosphere, reaching as high as 300 miles at noon in the summertime. Because it is the highest, it is the region mainly responsible for long-distance communications. A one-hop transmission can travel a maximum distance of about 2,500 miles using F2 propagation.

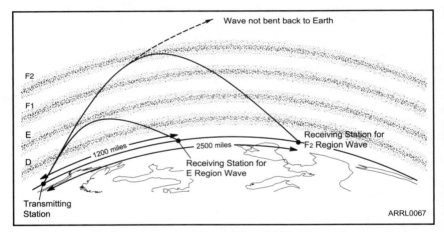

**Figure G3-4 — Radio waves are refracted (bent) in the ionosphere and may return to Earth. If the radio waves are refracted to Earth from the F2 region, they may return to Earth about 2500 miles from the transmitting station. If the radio waves are refracted to Earth from the E region, they may return to Earth about 1200 miles from the transmitting station.**

**G3B10** What is the approximate maximum distance along the Earth's surface that is normally covered in one hop using the E region?

A. 180 miles
B. 1,200 miles
C. 2,500 miles
D. 12,000 miles

**(B)** The E region of the ionosphere is the second lowest, just above the D region. The E layer forms at an altitude of about 70 miles above the Earth. The E region ionizes during the daytime, but does not stay ionized very long after sunset. Ionization in the E region is at a maximum around midday. During the daytime, a radio signal can travel a maximum distance of about 1,200 miles in one hop using E-region propagation.

**G3B11** What happens to HF propagation when the Lowest Usable Frequency (LUF) exceeds the Maximum Usable Frequency (MUF)?

A. No HF radio frequency will support ordinary skywave communications over the path
B. HF communications over the path are enhanced
C. Double hop propagation along the path is more common
D. Propagation over the path on all HF frequencies is enhanced

**(A)** Signals at frequencies below the LUF will be absorbed in the ionosphere rather than returning to Earth. Occasionally the LUF may be higher than the maximum usable frequency (MUF). This means that for the highest possible frequency that will propagate through the ionosphere for that path, the signal absorption is so large that even signals at the MUF are absorbed. Under these conditions it is impossible to establish sky-wave communication between those two points no matter what frequency is used! (Communications between either location and other locations may be possible, since the LUF and MUF depend on the end points of the communication path.)

**G3B12** **What factors affect the Maximum Usable Frequency (MUF)?**

A. Path distance and location
B. Time of day and season
C. Solar radiation and ionospheric disturbances
D. All of these choices are correct

**(D)** Maximum Usable Frequency (MUF) is the highest frequency that will provide sky-wave propagation between two specific locations. For example, suppose you live in Illinois and want to communicate with another amateur in Ecuador. You might find that the MUF for this contact is about 18 MHz at 14:00 UTC. You may also find that the MUF to communicate with a station in Spain at that same time is 12 MHz. Different communications path distances and directions will often result in very different MUF values. The MUF depends on conditions in the ionosphere, and those conditions will vary by time of day as well as the season of the year. The amount of solar radiation striking the ionosphere varies significantly depending on the timing of the 11-year sunspot cycle. Any solar flares, coronal-mass ejections and other disturbances on the Sun can also result in ionospheric disturbances that will affect the MUF. Answer choices A, B and C all describe factors that will affect the MUF for a given sky-wave propagation path, so answer choice D is correct.

## G3C    Ionospheric layers; critical angle and frequency; HF scatter; Near Vertical Incidence Sky waves

**G3C01** **Which of the following ionospheric layers is closest to the surface of the Earth?**

A. The D layer
B. The E layer
C. The F1 layer
D. The F2 layer

**(A)** The D region of the ionosphere is the lowest, forming the D layer at a height of 30 to 60 miles. Because it is the lowest, it is also the densest and its ionization disappears by dark.

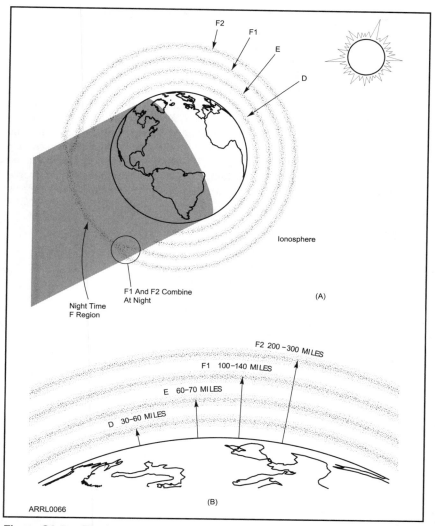

Figure G3-5 — The ionosphere consists of several regions of ionized particles at different heights above the Earth. At night, the D and E regions disappear and the F1 and F2 regions combine to form a single F region.

**G3C02** Where on the Earth do ionospheric layers reach their maximum height?

A. Where the Sun is overhead
B. Where the Sun is on the opposite side of the Earth
C. Where the Sun is rising
D. Where the Sun has just set

**(A)** The F region forms and decays in correlation with the daily passage of the Sun. The F1 and F2 layers form when the F region splits into two parts due to receiving high radiation from the Sun, recombining into a single F region at night. The more solar radiation the F region receives, the more it is ionized so it reaches maximum ionization shortly after noon during the summertime. The ionization tapers off very gradually towards sunset and the F2 region remains usable into the night. The F2 region is the highest of the ionosphere, reaching as high as 300 miles at noon in the summertime.

**G3C03** Why is the F2 region mainly responsible for the longest distance radio wave propagation?

A. Because it is the densest ionospheric layer
B. Because it does not absorb radio waves as much as other ionospheric regions
C. Because it is the highest ionospheric region
D. All of these choices are correct

**(C)** Because the F2 region is the highest ionospheric region, it is the region mainly responsible for long-distance communications. A one-hop transmission can travel a maximum distance of about 2500 miles using this F2 region.

**G3C04** What does the term "critical angle" mean as used in radio wave propagation?

A. The long path azimuth of a distant station
B. The short path azimuth of a distant station
C. The lowest takeoff angle that will return a radio wave to the Earth under specific ionospheric conditions
D. The highest takeoff angle that will return a radio wave to the Earth under specific ionospheric conditions

**(D)** At each frequency there is a maximum angle for which the radio wave can leave the antenna and still be refracted back to Earth by the ionosphere instead of simply passing through it and proceeding out into space. The critical angle changes depending on the ionization of the ionosphere.

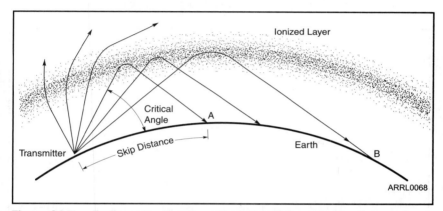

Figure G3-6 — Radio waves that leave the transmitting antenna at an angle higher than the critical angle are not refracted enough to return to Earth. A radio wave at the critical angle will return to Earth. The lowest-angle wave will return to Earth farther away than the wave at the critical angle. This illustrates the importance of low radiation angles for working DX.

**G3C05** Why is long distance communication on the 40, 60, 80 and 160 meter bands more difficult during the day?

  A. The F layer absorbs signals at these frequencies during daylight hours
  B. The F layer is unstable during daylight hours
  C. The D layer absorbs signals at these frequencies during daylight hours
  D. The E layer is unstable during daylight hours

**(C)** Think of the D region as the Darned Daylight region. Instead of bending high frequency signals back to Earth, it absorbs energy from them. Signals at lower frequencies (longer wavelengths such as 160, 80, 60 and 40 meters) are absorbed more than at higher frequencies. The ionization created by the sunlight does not last very long in the D region, disappearing by sunset.

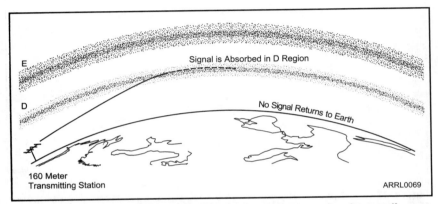

**Figure G3-7 — The D region of the ionosphere absorbs energy from radio waves. Lower-frequency radio waves don't make it all the way through the D region, so the waves do not return to Earth. Higher-frequency waves travel through the D region, and are then refracted (bent) back to Earth.**

**G3C06** **What is a characteristic of HF scatter signals?**

A. They have high intelligibility
B. They have a wavering sound
C. They have very large swings in signal strength
D. All of these choices are correct

**(B)** The area between the farthest reach of ground-wave propagation and the point where signals are refracted back from the ionosphere (sky-wave propagation) is called the skip zone. Since some of the transmitted signal is scattered in the atmosphere or from ground reflections, communication may be possible in the skip zone by the use of scatter signals. The amount of signal scattered in the atmosphere will be quite small and the signal received in the skip zone will arrive from several radio-wave paths. This tends to produce a weak, distorted signal with a fluttering or wavering sound.

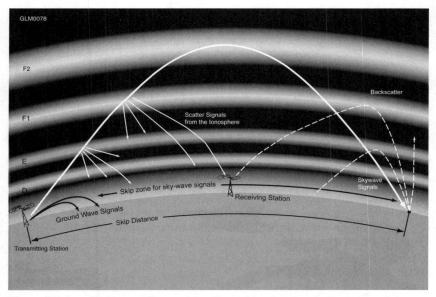

**Figure G3-8 — Radio waves may be reflected back towards the transmitting station from variations in the ionosphere or after striking the ground after ionospheric reflection. Some of this energy may be scattered back into the skip zone as a weak, highly variable signal.**

**G3C07**  What makes HF scatter signals often sound distorted?

A.  The ionospheric layer involved is unstable
B.  Ground waves are absorbing much of the signal
C.  The E-region is not present
D.  Energy is scattered into the skip zone through several different radio wave paths

**(D)** The amount of signal scattered back toward the transmitting station from the ionosphere or ground will be quite small. The signal received in the skip zone will also arrive from several radio-wave paths. This tends to produce a weak, distorted signal with a fluttering or wavering sound. (see also discussion for question G3C06)

**G3C08**  Why are HF scatter signals in the skip zone usually weak?

A.  Only a small part of the signal energy is scattered into the skip zone
B.  Signals are scattered from the magnetosphere which is not a good reflector
C.  Propagation is through ground waves which absorb most of the signal energy
D.  Propagation is through ducts in F region which absorb most of the energy

**(A)** The amount of signal scattered back toward the transmitting station from the ionosphere or ground will be quite small. The signal received in the skip zone will also arrive from several radio-wave paths. This tends to produce a weak, distorted signal with a fluttering or wavering sound. (see also discussion for question G3C06)

**G3C09**  What type of radio wave propagation allows a signal to be detected at a distance too far for ground wave propagation but too near for normal sky-wave propagation?

A.  Faraday rotation
B.  Scatter
C.  Sporadic-E skip
D.  Short-path skip

**(B)** The area between the farthest reach of ground-wave propagation and the point where signals are refracted back from the ionosphere (sky-wave propagation) is called the skip zone. Since some of the transmitted signal is scattered in the atmosphere, communication may be possible in the skip zone by the use of scatter signals. (see also discussion for question G3C06)

**G3C10** Which of the following might be an indication that signals heard on the HF bands are being received via scatter propagation?

A. The communication is during a sunspot maximum
B. The communication is during a sudden ionospheric disturbance
C. The signal is heard on a frequency below the Maximum Usable Frequency
D. The signal is heard on a frequency above the Maximum Usable Frequency

**(D)** Frequencies above the Maximum Usable Frequency (MUF) normally pass through the ionosphere out into space rather than being bent back, although atmospheric scatter from the ionosphere will sometimes allow communication on these frequencies. Amateurs trying to communicate on frequencies that seem to be above the MUF may notice that they can communicate using these scattered signals. Had they been using a frequency below the MUF they may not have noticed any scattered signals. (see also discussion for question G3C06)

**G3C11** Which of the following antenna types will be most effective for skip communications on 40 meters during the day?

A. Vertical antennas
B. Horizontal dipoles placed between ⅛ and ¼ wavelength above the ground
C. Left-hand circularly polarized antennas
D. Right-hand circularly polarized antenna

**(B)** Low horizontal antennas, such as dipoles between ⅛ and ¼ wavelength above the ground work best for daytime skip communications on low frequencies. Signals from such antennas are radiated at high vertical angles that can be reflected by the ionosphere, but have a minimum amount of attenuation from the D and E layers.

**G3C12** Which ionospheric layer is the most absorbent of long skip signals during daylight hours on frequencies below 10 MHz?

A. The F2 layer
B. The F1 layer
C. The E layer
D. The D layer

**(D)** See question G3C05.

**G3C13** What is Near Vertical Incidence Sky-wave (NVIS) propagation?

A. Propagation near the MUF
B. Short distance HF propagation using high elevation angles
C. Long path HF propagation at sunrise and sunset
D. Double hop propagation near the LUF

**(B)** Near Vertical Incidence Sky-wave (NVIS) propagation refers to communication using sky-wave signals transmitted at very high vertical angles. The frequencies used are below the critical frequency, meaning that their critical angle (see question G3C04) is ninety degrees, meaning they can be reflected straight back down to Earth. Because the signals travel at high angles, they have a minimum amount of attenuation from the D and E layers. The result is good communications in a region around the transmitter over distances higher than supported by ground-wave propagation.

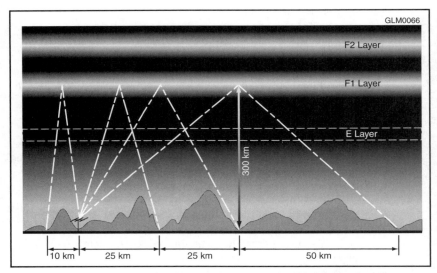

**Figure G3-9 — Near Vertical Incidence Sky-wave (NVIS) communications relies on signals below the critical frequency transmitted at high vertical angles. The signals are reflected by the ionosphere back to Earth in the region around the transmitter.**

# Amateur Radio Practices

Your General class exam (Element 3) will consist of 35 questions taken from the General class question pool as prepared by the Volunteer Examiner Coordinators' Questions Pool Committee. A certain number of questions are taken from each of the 10 subelements. There will be 5 questions from the subelement shown in this chapter. These questions are divided into 5 groups, labeled G4A through G4E.

## SUBELEMENT G4 — AMATEUR RADIO PRACTICES
## [5 Exam Questions — 5 Groups]

### G4A    Station Operation and Setup

**G4A01**    **What is the purpose of the "notch filter" found on many HF transceivers?**

A. To restrict the transmitter voice bandwidth
B. To reduce interference from carriers in the receiver passband
C. To eliminate receiver interference from impulse noise sources
D. To enhance the reception of a specific frequency on a crowded band

**(B)** A notch filter removes a very narrow band of frequencies — the "notch" — from its input. This allows the notch filter to get rid of an interfering tone, such as from an unmodulated carrier, while maintaining the intelligibility of the desired signal. An automatic notch filter can detect tones and remove them without operator intervention.

**G4A02**    **What is one advantage of selecting the opposite or "reverse" sideband when receiving CW signals on a typical HF transceiver?**

A. Interference from impulse noise will be eliminated
B. More stations can be accommodated within a given signal passband
C. It may be possible to reduce or eliminate interference from other signals
D. Accidental out of band operation can be prevented

**(C)** Interference from nearby signals can often be avoided by moving the receiver's carrier frequency to the "other side" of the desired signal, without changing its audio pitch. This won't work on SSB because it also inverts the spectrum of the speech, rendering it unintelligible.

**G4A03** What is normally meant by operating a transceiver in "split" mode?

A. The radio is operating at half power
B. The transceiver is operating from an external power source
C. The transceiver is set to different transmit and receive frequencies
D. The transmitter is emitting a SSB signal, as opposed to DSB operation

**(C)** Operating split means using one VFO for transmitting and another for receiving. This is often used when a rare DX station is on the air with many callers. Operating simplex — transmitting and receiving on the same frequency — can result in a lot of confusion with stations trying to hear the DX station while others transmit. By transmitting on one frequency and having callers transmit somewhere else (usually on adjacent frequencies) everyone can hear the DX station and keep "in sync" for more orderly, effective operating.

**G4A04** What reading on the plate current meter of a vacuum tube RF power amplifier indicates correct adjustment of the plate tuning control?

A. A pronounced peak
B. A pronounced dip
C. No change will be observed
D. A slow, rhythmic oscillation

**(B)** The Tune control of a vacuum tube RF amplifier — either of a standalone amplifier or of a transmitter output stage — adjusts the impedance matching circuit at the frequency of operation. Adjusting the Tune control for a pronounced "dip" in plate current indicates that the circuit is set for the right frequency. The Load or Coupling control is then used to adjust the amount of output power. The Tune and Load controls are alternately adjusted until the required amount of output power is obtained without exceeding the tube's plate current rating.

**G4A05** What is a purpose of using Automatic Level Control (ALC) with a RF power amplifier?

A. To balance the transmitter audio frequency response
B. To reduce harmonic radiation
C. To reduce distortion due to excessive drive
D. To increase overall efficiency

**(C)** The ALC circuit of an amplifier senses the amount of input power and generates a feedback voltage to keep a driver stage or transmitter from generating too much power for linear operation. This helps prevent distortion in the amplifier that would cause spurious emissions and interference to other stations.

**G4A06**  What type of device is often used to enable matching the transmitter output to an impedance other than 50 ohms?

A. Balanced modulator
B. SWR Bridge
C. Antenna coupler
D. Q Multiplier

**(C)** There are many names for devices that use LC circuits to convert one impedance to another; antenna coupler, impedance matching unit, transmatch and antenna tuner are common.

**G4A07**  What condition can lead to permanent damage when using a solid-state RF power amplifier?

A. Exceeding the Maximum Usable Frequency
B. Low input SWR
C. Shorting the input signal to ground
D. Excessive drive power

**(D)** Transistors are more sensitive to input drive levels than the more rugged vacuum tubes and can be damaged very quickly if too much power is applied. Fast-acting control circuits are required to protect the transistors from excessive drive.

**G4A08**  What is the correct adjustment for the load or coupling control of a vacuum tube RF power amplifier?

A. Minimum SWR on the antenna
B. Minimum plate current without exceeding maximum allowable grid current
C. Highest plate voltage while minimizing grid current
D. Maximum power output without exceeding maximum allowable plate current

**(D)** The Tune control of a vacuum tube RF amplifier — either of a standalone amplifier or of a transmitter output stage — adjusts the impedance matching circuit at the frequency of operation. Adjusting the Tune control for a pronounced "dip" in plate current indicates that the circuit is set for the right frequency. The Load or Coupling control is then used to adjust the amount of output power. The Tune and Load controls are alternately adjusted until the required amount of output power is obtained without exceeding the tube's plate current rating.

**G4A09** Why is a time delay sometimes included in a transmitter keying circuit?

A. To prevent stations from talking over each other
B. To allow the transmitter power regulators to charge properly
C. To allow time for transmit-receive changeover operations to complete properly before RF output is allowed
D. To allow time for a warning signal to be sent to other stations

**(C)** If a relay switches while RF is present in the circuit, that is called "hot switching". At high power levels hot switching can damage the relay and so it is important to let the relay complete switching before energizing the circuit. Similarly, it is important to disconnect sensitive receive circuits before transmitter RF is enabled. A controlled time delay, called "sequencing" is used to ensure that all relay and switching operations are completed before enabling the transmitter.

**G4A10** What is the purpose of an electronic keyer?

A. Automatic transmit/receive switching
B. Automatic generation of strings of dots and dashes for CW operation
C. VOX operation
D. Computer interface for PSK and RTTY operation

**(B)** Electronic keyers eliminate a lot of the manual work involved in operating a straight key and ensure that each dot or dash has the right length and spacing. This allows comfortable high-speed Morse operation even for extended periods of time.

**G4A11** Which of the following is a use for the IF shift control on a receiver?

A. To avoid interference from stations very close to the receive frequency
B. To change frequency rapidly
C. To permit listening on a different frequency from that on which you are transmitting
D. To tune in stations that are slightly off frequency without changing your transmit frequency

**(A)** Passband or IF shift adjusts the receiver's passband above or below the displayed carrier frequency to avoid interfering signals. This results in a shift in tone of the received signal, but often improves intelligibility.

**G4A12** Which of the following is a common use for the dual VFO feature on a transceiver?

A. To allow transmitting on two frequencies at once
B. To permit full duplex operation, that is transmitting and receiving at the same time
C. To permit ease of monitoring the transmit and receive frequencies when they are not the same
D. To facilitate computer interface

**(C)** Dual VFOs are used for split operation. (See the discussion for G4A03)

**G4A13** What is one reason to use the attenuator function that is present on many HF transceivers?

A. To reduce signal overload due to strong incoming signals
B. To reduce the transmitter power when driving a linear amplifier
C. To reduce power consumption when operating from batteries
D. To slow down received CW signals for better copy

**(A)** Too strong an input signal can overload a receiver, creating distortion products that interfere with reception of the desired signal. Using an attenuator reduces the input signal level and the possibility of overload.

**G4A14** How should the transceiver audio input be adjusted when transmitting PSK31 data signals?

A. So that the transceiver is at maximum rated output power
B. So that the transceiver ALC system does not activate
C. So that the transceiver operates at no more than 25% of rated power
D. So that the transceiver ALC indicator shows half scale

**(B)** For PSK31 operation, it is important that the output waveform be undistorted. By setting the transmitter output power so that the internal ALC circuit does not activate, that ensures the output amplifier is operating with maximum linearity.

## G4B  Test and monitoring equipment; two-tone test

**G4B01** What item of test equipment contains horizontal and vertical channel amplifiers?

A. An ohmmeter
B. A signal generator
C. An ammeter
D. An oscilloscope

**(D)** An oscilloscope uses one signal to deflect a beam of electrons horizontally across the face of the screen and a second signal to deflect them vertically. The electron beam causes a phosphor coating on the inside of the screen to glow so that the beam's position can be observed. The horizontal and vertical channel amplifiers in the oscilloscope increase the voltage of the two input signals for the desired amount of deflection. Most oscilloscopes have a built-in oscillator circuit, called a sweep generator, designed to repeatedly deflect the electron beam horizontally across the screen at a fixed rate. With an external signal applied to the vertical amplifier, the sweep generator frequency is adjusted to produce a stable pattern on the screen. Some oscilloscopes also allow a second input signal to be applied to a separate horizontal channel amplifier. This allows two signals to be compared to each other with respect to time.

**G4B02** Which of the following is an advantage of an oscilloscope versus a digital voltmeter?

A. An oscilloscope uses less power
B. Complex impedances can be easily measured
C. Input impedance is much lower
D. Complex waveforms can be measured

**(D)** Digital voltmeters display measured values of voltage with excellent precision but they cannot measure a complex signal waveform's time-related parameters, such as its frequency, period, or how it reacts to other signals.

**G4B03** Which of the following is the best instrument to use when checking the keying waveform of a CW transmitter?

A. An oscilloscope
B. A field-strength meter
C. A sidetone monitor
D. A wavemeter

**(A)** An oscilloscope visually displays a signal waveform. This allows you to observe the shape of the CW signal (referred to as the CW envelope) noting, for example, the rise and fall time of the signal. It also allows you to observe problems such as flat-topping (caused by overmodulation) on your SSB signal.

**G4B04** What signal source is connected to the vertical input of an oscilloscope when checking the RF envelope pattern of a transmitted signal?

A. The local oscillator of the transmitter
B. An external RF oscillator
C. The transmitter balanced mixer output
D. The attenuated RF output of the transmitter

**(D)** When the RF output from a transmitter is connected to the vertical channel of an oscilloscope, the oscilloscope visually displays the signal's envelope. This allows you to check for signal distortion such as flat-topping (caused by overmodulation). Use an attenuator or RF sampling device to limit the voltage at the oscilloscope input.

**G4B05** Why is high input impedance desirable for a voltmeter?

A. It improves the frequency response
B. It decreases battery consumption in the meter
C. It improves the resolution of the readings
D. It decreases the loading on circuits being measured

**(D)** The higher the impedance of a voltmeter, the smaller the amount of current drawn from the circuit being tested. This allows the voltmeter to make an accurate measurement of voltage while disturbing the circuit as little as possible.

**G4B06** What is an advantage of a digital voltmeter as compared to an analog voltmeter?

A. Better for measuring computer circuits
B. Better for RF measurements
C. Better precision for most uses
D. Faster response

(C) A digital voltmeter displays measurements of voltage, current and resistance in numeric form instead of using a moving needle and a fixed scale. That results in significantly better precision than an analog meter. Analog meters, however, may be a better choice for adjusting a circuit for peak and minimum values because the needle movement is easier to see than it is to read changes on a numeric display.

**G4B07** Which of the following might be a use for a field strength meter?

A. Close-in radio direction-finding
B. A modulation monitor for a frequency or phase modulation transmitter
C. An overmodulation indicator for a SSB transmitter
D. A keying indicator for a RTTY or packet transmitter

(A) A field-strength meter makes a relative measurement of the intensity of the field being radiated from an antenna. If you are doing some radio direction finding (RDF), you may find that a field-strength meter is a handy piece of equipment to use as you get close to the transmitter source. A strong signal may drive the S meter off scale, requiring a variable attenuator to be used between the directional antenna and receiver as you get closer to the transmitter. A field-strength meter could replace the normal RDF equipment under those conditions, and you can use your body to shield the field-strength meter to indicate the direction to the transmitter.

**G4B08** Which of the following instruments may be used to monitor relative RF output when making antenna and transmitter adjustments?

A. A field-strength meter
B. An antenna noise bridge
C. A multimeter
D. A Q meter

(A) A field-strength meter makes a relative measurement of the intensity of the field being radiated from an antenna. Some amateurs keep a field-strength meter in their shack to monitor the relative RF output from their station. This can be especially handy when you are making antenna or transmitter adjustments

**G4B09** Which of the following can be determined with a field strength meter?

A. The radiation resistance of an antenna
B. The radiation pattern of an antenna
C. The presence and amount of phase distortion of a transmitter
D. The presence and amount of amplitude distortion of a transmitter

**(B)** A field-strength meter makes a relative measurement of the intensity of the field being radiated from an antenna. By placing the field-strength meter in different locations around the antenna, you can determine the relative field pattern of the antenna.

**G4B10** Which of the following can be determined with a directional wattmeter?

A. Standing wave ratio
B. Antenna front-to-back ratio
C. RF interference
D. Radio wave propagation

**(A)** SWR can be calculated from forward and reflected power measurements made using a directional wattmeter. SWR is then calculated using the following formula:

$$SWR = \frac{1 + \sqrt{P_R / P_F}}{1 - \sqrt{P_R / P_F}}$$

where $P_F$ is forward power and $P_R$ is reflected power.

**G4B11** Which of the following must be connected to an antenna analyzer when it is being used for SWR measurements?

A. Receiver
B. Transmitter
C. Antenna and feed line
D. All of these choices are correct

**(C)** An antenna analyzer is the equivalent of a very low-power, adjustable-frequency transmitter and SWR bridge. The antenna and feed line are connected to the analyzer and SWR measurements are made directly from the analyzer's meter or display while the analyzer frequency is adjusted. This is much more convenient than using a transmitter and wattmeter and also minimizes the potential for interfering with other signals.

**G4B12** What problem can occur when making measurements on an antenna system with an antenna analyzer?

A. SWR readings may be incorrect if the antenna is too close to the Earth

B. Strong signals from nearby transmitters can affect the accuracy of measurements

C. The analyzer can be damaged if measurements outside the ham bands are attempted

D. Connecting the analyzer to an antenna can cause it to absorb harmonics

**(B)** An antenna analyzer is the equivalent of a very low-power, adjustable-frequency transmitter and SWR bridge. Because the SWR bridge must be sensitive enough to work with the low-power transmitter, it is also sensitive to RF that the antenna may pick up. This is a particular problem when using the analyzer near broadcast stations with their high-powered transmitters. Symptoms might include SWR readings that change with station programming and excessively high or low SWR that does not change with frequency as expected.

**G4B13** What is a use for an antenna analyzer other than measuring the SWR of an antenna system?

A. Measuring the front to back ratio of an antenna

B. Measuring the turns ratio of a power transformer

C. Determining the impedance of an unknown or unmarked coaxial cable

D. Determining the gain of a directional antenna

**(C)** An antenna analyzer's manual will show how to make many useful measurements such as feed line characteristic impedance, velocity of propagation, electrical length, and so on. These are very flexible test instruments.

**G4B14** What is an instance in which the use of an instrument with analog readout may be preferred over an instrument with a numerical digital readout?

A. When testing logic circuits

B. When high precision is desired

C. When measuring the frequency of an oscillator

D. When adjusting tuned circuits

**(D)** The analog meter's moving needle across calibrated scales on the meter face is much easier to adjust for a maximum or minimum value than a numeric display.

**G4B15** What type of transmitter performance does a two-tone test analyze?

A. Linearity
B. Carrier and undesired sideband suppression
C. Percentage of frequency modulation
D. Percentage of carrier phase shift

**(A)** It is common to test the amplitude linearity of a single-sideband transmitter by injecting two audio tones of equal level into the microphone jack, then observing the pattern made on an oscilloscope. In order to get meaningful results, the two tones must not be harmonically related to each other (such as 1 and 2 kHz). Of course, in order for the audio frequencies to display on the oscilloscope, they must be within the audio passband of the transmitter. The ARRL Lab uses 700 Hz and 1900 Hz tones to perform this test.

**G4B16** What signals are used to conduct a two-tone test?

A. Two audio signals of the same frequency shifted 90-degrees
B. Two non-harmonically related audio signals
C. Two swept frequency tones
D. Two audio frequency range square wave signals of equal amplitude

**(B)** See question G4B15

## G4C Interference with consumer electronics; grounding; DSP

**G4C01** Which of the following might be useful in reducing RF interference to audio-frequency devices?

A. Bypass inductor
B. Bypass capacitor
C. Forward-biased diode
D. Reverse-biased diode

**(B)** If radio frequency interference is entering a home audio system through external control cables or power leads, a bypass capacitor can be effective at keeping the unwanted RF signal out of the equipment. With transistor or integrated circuit audio amplifiers you may need to use RF chokes in series with the speaker leads instead of a bypass capacitor. See the *ARRL Handbook* and *The ARRL RFI Book* for more information on finding and fixing RFI problems.

**G4C02**  Which of the following could be a cause of interference covering a wide range of frequencies?

A. Not using a balun or line isolator to feed balanced antennas
B. Lack of rectification of the transmitter's signal in power conductors
C. Arcing at a poor electrical connection
D. The use of horizontal rather than vertical antennas

**(C)** An arc, such as in motors or at the contacts of electrical equipment, is rich in harmonic energy, even though the primary current may be dc or 60 Hz ac. The resulting RF harmonics can be radiated by the power wiring as broadband noise heard by nearby receivers. Broadband noise can also be caused by intermittent or poor contacts in RF grounds, such as those in your own station.

**G4C03**  What sound is heard from an audio device or telephone if there is interference from a nearby single-sideband phone transmitter?

A. A steady hum whenever the transmitter is on the air
B. On-and-off humming or clicking
C. Distorted speech
D. Clearly audible speech

**(C)** An audio device or telephone can sometimes rectify and detect RF signals in much the same way that a radio does. The audio signal is then amplified resulting in interference. The amateur's voice will be heard but it will be highly distorted.

**G4C04**  What is the effect on an audio device or telephone system if there is interference from a nearby CW transmitter?

A. On-and-off humming or clicking
B. A CW signal at a nearly pure audio frequency
C. A chirpy CW signal
D. Severely distorted audio

**(A)** An audio device or telephone can sometimes rectify and detect RF signals in much the same way that a radio does. The audio signal is then amplified resulting in interference. An amateur's CW transmission will be heard as on-and-off humming or clicking.

**G4C05** What might be the problem if you receive an RF burn when touching your equipment while transmitting on an HF band, assuming the equipment is connected to a ground rod?

A. Flat braid rather than round wire has been used for the ground wire
B. Insulated wire has been used for the ground wire
C. The ground rod is resonant
D. The ground wire has high impedance on that frequency

**(D)** One purpose of an amateur station's RF ground system is to make sure that unwanted signals, such as those picked up from your own transmissions, are directed to ground instead of flowing on or between pieces of equipment. If the ground wire is long enough to be resonant on one or more bands, however, it will present a high impedance, which can enable high RF voltages to be present on the chassis of your equipment or microphone. This can cause an RF burn if the "hot spot" is touched when you are transmitting.

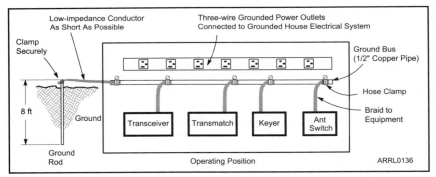

**Figure G4-1 — You can make an effective station ground by connecting all equipment to a ground bus. A length of ½ inch copper pipe along the back of your operating desk or table makes a good ground bus. Heavy copper braid, such as the outer braid from RG-8 coaxial cable makes a good, flexible strap to connect each piece of equipment to the ground bus. The whole system then connects to a good earth ground — an 8-foot ground rod located as close to the station as possible is the minimum ground you should use. More than one ground rod may be needed in some locations.**

**G4C06** What effect can be caused by a resonant ground connection?

A. Overheating of ground straps
B. Corrosion of the ground rod
C. High RF voltages on the enclosures of station equipment
D. A ground loop

**(C)** See question G4C05

**G4C07** What is one good way to avoid unwanted effects of stray RF energy in an amateur station?

A. Connect all equipment grounds together
B. Install an RF filter in series with the ground wire
C. Use a ground loop for best conductivity
D. Install a few ferrite beads on the ground wire where it connects to your station

**(A)** By keeping all of your equipment at the same potential, RF "hot spots" (high voltages) won't be present and RF current that can cause distortion or improper operation won't flow between the enclosures. This is a particularly good strategy when the connection to your ground rod is long enough to be resonant on one or more bands.

**G4C08** Which of the following would reduce RF interference caused by common-mode current on an audio cable?

A. Placing a ferrite bead around the cable
B. Adding series capacitors to the conductors
C. Adding shunt inductors to the conductors
D. Adding an additional insulating jacket to the cable

**(A)** The best solution to many types of interference caused by proximity to an amateur station is to keep the RF signals from entering the equipment in the first place. If filters can be used, they are generally the most effective and least troublesome to install. The next approach is to prevent RF current flow by placing inductance or resistance in its path. This is done by forming the conductor carrying the RF current into an RF choke by winding it around or through a ferrite core. Ferrite beads and cores can also be placed on cables to prevent RF common-mode current from flowing on the outside of cable braids or shields ("common-mode" interference).

**G4C09** How can a ground loop be avoided?

A. Connect all ground conductors in series
B. Connect the AC neutral conductor to the ground wire
C. Avoid using lock washers and star washers when making ground connections
D. Connect all ground conductors to a single point

**(D)** Ground loops are created when a continuous current path (the loop) exists around a series of equipment enclosures. This loop acts as a single-turn inductor that picks up voltages from magnetic fields generated by power transformers, ac wiring and other low-frequency currents. The result is a "hum" or "buzz" in audio signals or an ac signal that interferes with control or data signals. Less frequently, the loop can pick up transmitted RF and cause distortion in audio signals. Ground loops can be avoided by using a star-ground or ground bus system where all grounds share a common, low-impedance connection. If interconnections make it impossible to avoid loops, minimize the loop's area and inductance by bundling cables together.

**G4C10** **What could be a symptom of a ground loop somewhere in your station?**

A. You receive reports of "hum" on your station's transmitted signal
B. The SWR reading for one or more antennas is suddenly very high
C. An item of station equipment starts to draw excessive amounts of current
D. You receive reports of harmonic interference from your station

**(A)** A ground loop can pick up magnetic fields from transformers or ac wiring that result in a "hum" or "buzz" in audio signals or an ac signal that interferes with control or data signals. Less frequently, the loop can pick up transmitted RF and cause distortion in audio signals.

**G4C11** **Which of the following is one use for a Digital Signal Processor in an amateur station?**

A. To provide adequate grounding
B. To remove noise from received signals
C. To increase antenna gain
D. To increase antenna bandwidth

**(B)** Digital Signal Processing (DSP) is the process of converting analog signals to digital data, processing them with software programs, then converting the signals back to analog form. Digital noise reduction is but one application — filtering, demodulation and decoding are also commonly performed by DSP circuits.

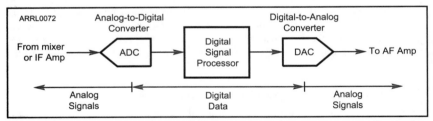

**Figure G4-2 — A DSP circuit consists of an analog-to-digital converter (ADC), a microprocessor that operates on the digitized data, and a digital-to-analog converter (DAC) that converts the signal back to analog.**

**G4C12** Which of the following is an advantage of a receiver Digital Signal Processor IF filter as compared to an analog filter?

A. A wide range of filter bandwidths and shapes can be created
B. Fewer digital components are required
C. Mixing products are greatly reduced
D. The DSP filter is much more effective at VHF frequencies

**(A)** Because DSP circuits use software to process the digitized signal, the number of different functions that can be performed is only limited by the amount of memory and processing speed. In addition, unlike most analog filters, parameters such as bandwidth and the shape of the response can be made adjustable by the operator. This provides a great deal of flexibility in receiver operation.

**G4C13** Which of the following can perform automatic notching of interfering carriers?

A. Band-pass tuning
B. A Digital Signal Processor (DSP) filter
C. Balanced mixing
D. A noise limiter

**(B)** The software program that controls a DSP notch filter may also operate automatically. In this type of operation, the software searches for the steady tones of an interfering carrier, such as from a shortwave broadcast signal, and then adjusts the frequency of the notch until the amplitude of the carrier's tone is minimized. (see also the discussion for question G1A01)

## G4D Speech processors; S meters; sideband operation near band edges

**G4D01** What is the purpose of a speech processor as used in a modern transceiver?

A. Increase the intelligibility of transmitted phone signals during poor conditions
B. Increase transmitter bass response for more natural sounding SSB signals
C. Prevent distortion of voice signals
D. Decrease high-frequency voice output to prevent out of band operation

**(A)** A speech processor can improve signal intelligibility by raising average power without increasing peak envelope power (PEP). It does this by bringing up low signal levels while not increasing high signal levels. As a result, the average signal level is higher. A speech processor does not increase the transmitter output PEP.

**G4D02** Which of the following describes how a speech processor affects a transmitted single sideband phone signal?

A. It increases peak power
B. It increases average power
C. It reduces harmonic distortion
D. It reduces intermodulation distortion

**(B)** A speech processor can improve signal intelligibility by raising average power without increasing peak envelope power (PEP). It does this by bringing up low signal levels while not increasing high signal levels. As a result, the average signal level is higher.

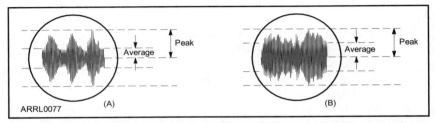

ARRL0077 (A) (B)

**Figure G4-3 — A typical SSB voice-modulated signal might have an envelope similar to the oscilloscope display shown at A. The RF amplitude (current or voltage) is on the vertical axis, and the time sweep is on the horizontal axis. After speech processing, the envelope pattern might look like the display at B. The average power of the processed signal has increased, but the PEP has not changed.**

**G4D03** Which of the following can be the result of an incorrectly adjusted speech processor?

A. Distorted speech
B. Splatter
C. Excessive background pickup
D. All of these choices are correct

**(D)** Proper adjustment of a speech processor is important to ensure your transmitted signal is not distorted and is free of spurious signals. Increasing the amount of processing too far causes your speech to be distorted. Processors that amplify softer speech components more than loud ones will also pick up background noise from fans and other radios and combine them with your desired speech. Another common result of too much processing is overdriving the transmitter output stages, causing interference ("splatter") to signals on nearby frequencies. Read the owner's manual for your radio and learn to operate its speech processor correctly.

**G4D04**  What does an S meter measure?

A. Conductance
B. Impedance
C. Received signal strength
D. Transmitter power output

**(C)** An S meter measures received signal strength in S units.

**G4D05**  How does an S meter reading of 20 dB over S-9 compare to an S-9 signal, assuming a properly calibrated S meter?

A. It is 10 times weaker
B. It is 20 times weaker
C. It is 20 times stronger
D. It is 100 times stronger

**(D)** An ideal S meter operates on a logarithmic scale, indicating one S unit of change for a four-times increase or decrease in power. (This is a 6-dB change in power.) In real life, S meters are only calibrated to this standard in the middle of their range (if at all). Above S9, theoretically corresponding to a 50 μV input signal, S meters are calibrated in dB. In the example of this question, a signal 20 dB stronger than S9 is 100 times stronger than S9.

**G4D06**  Where is an S meter found?

A. In a receiver
B. In an SWR bridge
C. In a transmitter
D. In a conductance bridge

**(A)** An S meter measures received signal strength and so is found in a receiver or the receive circuits of a transceiver.

**G4D07**  How much must the power output of a transmitter be raised to change the S-meter reading on a distant receiver from S8 to S9?

A. Approximately 1.5 times
B. Approximately 2 times
C. Approximately 4 times
D. Approximately 8 times

**(C)** An ideal S meter operates on a logarithmic scale, indicating one S unit of change for a four-times increase or decrease in power. This is a 6 dB change in power. (See also the discussion for question G4D05.)

**G4D08** What frequency range is occupied by a 3 kHz LSB signal when the displayed carrier frequency is set to 7.178 MHz?

A. 7.178 to 7.181 MHz
B. 7.178 to 7.184 MHz
C. 7.175 to 7.178 MHz
D. 7.1765 to 7.1795 MHz

**(C)** Nearly all radios display the carrier frequency of a SSB transmission. That means your actual signal lies entirely above (USB) or below (LSB) the displayed frequency. If the sidebands occupy 3 kHz of spectrum, you'll need to stay far enough from the edge of your frequency privileges to avoid transmitting a signal outside them. For example, Generals are permitted to use up to 14.350 MHz, so the displayed carrier frequency of a USB signal should be no less than 3 kHz from the band edge — 14.347 MHz — and the signal occupies 14.347 to 14.350 MHz. If you transmit higher than that, the sidebands begin to extend into the non-amateur frequencies above 14.350 MHz! Similarly, using LSB on 40 meters, Generals should operate with the carrier frequency no less than 3 kHz from the band edge — 7.178 MHz — thus occupying the range of 7.175 to 7.178 MHz.

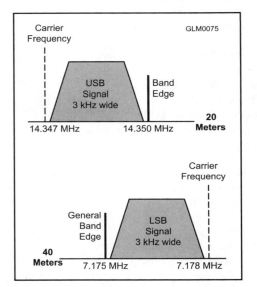

Figure G4-4 — When sidebands extend from the carrier towards a band edge or a band segment edge, operate with a displayed carrier frequency no closer than 3 kHz to the edge frequency and be sure your signal is "clean."

**G4D09** What frequency range is occupied by a 3 kHz USB signal with the displayed carrier frequency set to 14.347 MHz?

A. 14.347 to 14.647 MHz
B. 14.347 to 14.350 MHz
C. 14.344 to 14.347 MHz
D. 14.3455 to 14.3485 MHz

**(B)** See the discussion for question G4D08.

**G4D10** How close to the lower edge of the 40 meter General Class phone segment should your displayed carrier frequency be when using 3 kHz wide LSB?

A. 3 kHz above the edge of the segment
B. 3 kHz below the edge of the segment
C. Your displayed carrier frequency may be set at the edge of the segment
D. Center your signal on the edge of the segment

**(A)** See the discussion for question G4D08.

**G4D11** How close to the upper edge of the 20 meter General Class band should your displayed carrier frequency be when using 3 kHz wide USB?

A. 3 kHz above the edge of the band
B. 3 kHz below the edge of the band
C. Your displayed carrier frequency may be set at the edge of the band
D. Center your signal on the edge of the band

**(B)** See the discussion for question G4D08.

## G4E HF mobile radio installations; emergency and battery powered operation

**G4E01** What is a "capacitance hat", when referring to a mobile antenna?

A. A device to increase the power handling capacity of a mobile whip antenna
B. A device that allows automatic band-changing for a mobile antenna
C. A device to electrically lengthen a physically short antenna
D. A device that allows remote tuning of a mobile antenna

**(C)** To increase the capacitance of a mobile antenna and electrically lengthen it, a structure of rods and possibly a ring is added at or near the top of the antenna. This is called a "capacitance hat" or "capacity hat".

**G4E02** What is the purpose of a "corona ball" on a HF mobile antenna?

A. To narrow the operating bandwidth of the antenna
B. To increase the "Q" of the antenna
C. To reduce the chance of damage if the antenna should strike an object
D. To reduce high voltage discharge from the tip of the antenna

**(D)** The sharp tip of mobile whip can cause corona discharge even at moderate power levels. By adding a smooth ball, the tendency for corona to form is reduced.

**G4E03** Which of the following direct, fused power connections would be the best for a 100-watt HF mobile installation?

A. To the battery using heavy gauge wire
B. To the alternator or generator using heavy gauge wire
C. To the battery using resistor wire
D. To the alternator or generator using resistor wire

**(A)** When you are making the power connections for your 100-watt HF radio for mobile operation, you should connect heavy-gauge wires directly to the battery terminals. Remember that both leads should have fuses, placed as close to the battery as possible.

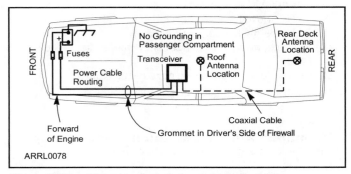

**Figure G4-5 — This drawing shows a typical mobile-transceiver-installation wiring diagram.**

**G4E04** Why is it best NOT to draw the DC power for a 100-watt HF transceiver from an automobile's auxiliary power socket?
A. The socket is not wired with an RF-shielded power cable
B. The socket's wiring may be inadequate for the current being drawn by the transceiver
C. The DC polarity of the socket is reversed from the polarity of modern HF
D. Drawing more than 50 watts from this socket could cause the engine to overheat

**(B)** The auxiliary socket wiring is adequate for a low-power hand-held radio but probably not for a full-power HF rig drawing 20 amps or more when transmitting. Use a direct connection to the battery. (See the discussion for G4E03)

**G4E05** Which of the following most limits the effectiveness of an HF mobile transceiver operating in the 75 meter band?
A. "Picket Fencing" signal variation
B. The wire gauge of the DC power line to the transceiver
C. The antenna system
D. FCC rules limiting mobile output power on the 75 meter band

**(C)** HF mobile antenna systems are the most limiting factor for effective operation of your station. Placing even an 8-foot vertical antenna on top of a small car makes a dangerously tall system. If you have a taller vehicle, this kind of antenna is almost out of the question. Some type of inductive loading to shorten the length is then required. For operation on the 75 meter band this compromise results in a relatively inefficient antenna system.

**G4E06** What is one disadvantage of using a shortened mobile antenna as opposed to a full size antenna?
A. Short antennas are more likely to cause distortion of transmitted signals
B. Short antennas can only receive vertically polarized signals
C. Operating bandwidth may be very limited
D. Harmonic radiation may increase

**(C)** Electrically short antennas have a very low impedance at the feed point. This causes the frequency range to be very narrow over which a matching network presents a 50-ohm impedance to the transmitter.

**G4E07** Which of the following is the most likely to cause interfering signals to be heard in the receiver of an HF mobile installation in a recent model vehicle?
A. The battery charging system
B. The anti-lock braking system
C. The anti-theft circuitry
D. The vehicle control computer

**(D)** Digital signals internal and external to the vehicle's control computer system can radiate from the vehicle's electrical wiring and be picked up by a mobile radio.

**G4E08** What is the name of the process by which sunlight is changed directly into electricity?

A. Photovoltaic conversion
B. Photon emission
C. Photosynthesis
D. Photon decomposition

**(A)** Natural sources of energy are becoming more economical and more practical as a way to power your station. Photovoltaic conversion by solar cells and panels converts sunlight into dc electricity.

**G4E09** What is the approximate open-circuit voltage from a modern, well-illuminated photovoltaic cell?

A. 0.02 VDC
B. 0.5 VDC
C. 0.2 VDC
D. 1.38 VDC

**(B)** Each photovoltaic cell produces about 0.5 volt in full sunlight if there is no load connected to the cell. The size or surface area of the cell determines the maximum current that the cell can supply.

**G4E10** What is the reason a series diode is connected between a solar panel and a storage battery that is being charged by the panel?

A. The diode serves to regulate the charging voltage to prevent overcharge
B. The diode prevents self discharge of the battery though the panel during times of low or no illumination
C. The diode limits the current flowing from the panel to a safe value
D. The diode greatly increases the efficiency during times of high illumination

**(B)** If connected directly to a battery, during periods of low or no illumination, the battery voltage will be higher than that from the panel, allowing the battery to discharge back through the panel.

**G4E11** Which of the following is a disadvantage of using wind as the primary source of power for an emergency station?

A. The conversion efficiency from mechanical energy to electrical energy is less than 2 percent

B. The voltage and current ratings of such systems are not compatible with amateur equipment

C. A large energy storage system is needed to supply power when the wind is not blowing

D. All of these choices are correct

**(C)** Wind-powered systems provide free energy after the initial cost of installation, but just like solar-powered systems, wind systems need to have a storage battery to supply electricity when the wind is not blowing.

# Electrical Principles

Your General class exam (Element 3) will consist of 35 questions taken from the General class question pool as prepared by the Volunteer Examiner Coordinators' Questions Pool Committee. A certain number of questions are taken from each of the 10 subelements. There will be 3 questions from the subelement shown in this chapter. These questions are divided into 3 groups, labeled G5A through G5C.

## SUBELEMENT G5 — ELECTRICAL PRINCIPLES
## [3 Exam Questions — 3 Groups]

### G5A    Reactance; inductance; capacitance; impedance; impedance matching

**G5A01**   What is impedance?
- A.  The electric charge stored by a capacitor
- B.  The inverse of resistance
- C.  The opposition to the flow of current in an AC circuit
- D.  The force of repulsion between two similar electric fields

**(C)**  Impedance is the opposition to the flow of current in an alternating current (ac) circuit made up of capacitive reactance, inductive reactance and resistance.

**G5A02**   What is reactance?
- A.  Opposition to the flow of direct current caused by resistance
- B.  Opposition to the flow of alternating current caused by capacitance or inductance
- C.  A property of ideal resistors in AC circuits
- D.  A large spark produced at switch contacts when an inductor is de-energized

**(B)**  The opposition to flow of ac current caused by inductance and capacitance is referred to as reactance. Reactance is one component of impedance, along with resistance.

**G5A03** Which of the following causes opposition to the flow of alternating current in an inductor?

A. Conductance
B. Reluctance
C. Admittance
D. Reactance

**(D)** The opposition to flow of ac current caused by inductance and capacitance is referred to as reactance. The opposition to flow of current in an alternating current (ac) circuit caused by an inductor is referred to as inductive reactance.

**G5A04** Which of the following causes opposition to the flow of alternating current in a capacitor?

A. Conductance
B. Reluctance
C. Reactance
D. Admittance

**(C)** The opposition to flow of ac current caused by inductance and capacitance is referred to as reactance. The opposition to flow of current in an alternating current (ac) circuit caused by a capacitor is referred to as capacitive reactance.

**G5A05** How does an inductor react to AC?

A. As the frequency of the applied AC increases, the reactance decreases
B. As the amplitude of the applied AC increases, the reactance increases
C. As the amplitude of the applied AC increases, the reactance decreases
D. As the frequency of the applied AC increases, the reactance increases

**(D)** The opposition to flow of current caused by the coil in an alternating current (ac) circuit is referred to as inductive reactance. Inductive reactance increases as the ac frequency increases.

**G5A06** How does a capacitor react to AC?

A. As the frequency of the applied AC increases, the reactance decreases
B. As the frequency of the applied AC increases, the reactance increases
C. As the amplitude of the applied AC increases, the reactance increases
D. As the amplitude of the applied AC increases, the reactance decreases

**(A)** The opposition to flow of current caused by a capacitor in an alternating current (ac) circuit is referred to as capacitive reactance. This reactance decreases as the ac frequency increases.

**G5A07**  What happens when the impedance of an electrical load is equal to the internal impedance of the power source?

A. The source delivers minimum power to the load
B. The electrical load is shorted
C. No current can flow through the circuit
D. The source can deliver maximum power to the load

**(D)** A power source delivers maximum power to a load when the impedance of the load is equal to (matched to) the impedance of the source. When the impedances are not matched, the power source cannot transfer as much power to the load. This is true for both dc power sources (such as batteries or power supplies) and ac power sources (such as transmitters).

**G5A08**  Why is impedance matching important?

A. So the source can deliver maximum power to the load
B. So the load will draw minimum power from the source
C. To ensure that there is less resistance than reactance in the circuit
D. To ensure that the resistance and reactance in the circuit are equal

**(A)** See the discussion for question G5A07.

**G5A09**  What unit is used to measure reactance?

A. Farad
B. Ohm
C. Ampere
D. Siemens

**(B)** The ohm is the unit used to measure any opposition to the flow of current. In an ac circuit, this opposition is referred to as impedance which includes both reactance and resistance.

**G5A10**  What unit is used to measure impedance?

A. Volt
B. Ohm
C. Ampere
D. Watt

**(B)** The ohm is the unit used to measure any opposition to the flow of current. In an ac circuit, this opposition is referred to as impedance which includes both reactance and resistance.

**G5A11** Which of the following describes one method of impedance matching between two AC circuits?

A. Insert an LC network between the two circuits
B. Reduce the power output of the first circuit
C. Increase the power output of the first circuit
D. Insert a circulator between the two circuits

**(A)** An LC network, such as a pi-network (see question G5A13), uses the exchange of stored energy between the inductor and capacitors to transform the ratio of voltage and current (impedance) between the input and output. Common examples of impedance matching LC networks are the L-, T-, and pi-network, named for the resemblance to a letter of the arrangement of their components on a schematic.

**G5A12** What is one reason to use an impedance matching transformer?

A. To minimize transmitter power output
B. To maximize the transfer of power
C. To reduce power supply ripple
D. To minimize radiation resistance

**(B)** An impedance matching transformer changes the ratio of voltage and current between the load and source. Since impedance is the ratio of voltage and current, the transformer can also match different impedances. Matching impedances also maximizes power transfer between the source and load.

**G5A13** Which of the following devices can be used for impedance matching at radio frequencies?

A. A transformer
B. A Pi-network
C. A length of transmission line
D. All of these choices are correct

**(D)** All of these can alter the ratio of voltage and current in a circuit, effectively changing the impedance as well. A pi-network is an LC-circuit that uses the exchange of stored energy between the inductor and capacitors to transform the ratio of voltage and current (impedance) between the input and output. Special lengths of transmission lines (called "stubs") set up patterns of reflections in the feed line that cancel the reflections from a mismatched load, making the load impedance appear as if it was the same as that of the feed line.

## G5B The Decibel; current and voltage dividers; electrical power calculations; sine wave root-mean-square (RMS) values; PEP calculations

**G5B01** A two-times increase or decrease in power results in a change of how many dB?

A. Approximately 2 dB
B. Approximately 3 dB
C. Approximately 6 dB
D. Approximately 12 dB

**(B)** The decibel scale is a logarithmic scale in which a two-times increase (or decrease) in power is represented by 3 dB. The mathematical formula for the decibel scale for power is:

$$dB = 10 \times \log_{10}\left(\frac{P_2}{P_1}\right)$$

where $P_1$ = reference power and $P_2$ = power being compared to the reference value.

In this case:

$$dB = 10 \times \log_{10}\left(\frac{2}{1}\right) = 10 \times \log_{10}(2) = 10 \times 0.3 = 3\ dB$$

---

**Table G5B01**

**Some Common Decibel Values and Power Ratio Equivalents**

| dB | P2 / P1 |
|---|---|
| 20 | 100 ($10^2$) |
| 10 | 10 ($10^1$) |
| 6 | 4 |
| 3 | 2 |
| 0 | 1 |
| −3 | 0.5 |
| −6 | 0.25 |
| −10 | 0.1 ($10^{-1}$) |
| −20 | 0.01 ($10^{-2}$) |

---

**G5B02** How does the total current relate to the individual currents in each branch of a parallel circuit?

A. It equals the average of each branch current
B. It decreases as more parallel branches are added to the circuit
C. It equals the sum of the currents through each branch
D. It is the sum of the reciprocal of each individual voltage drop

**(C)** In a circuit with several parallel branches, the total current flowing into the junction of the branches is equal to the sum of the current through each branch. This is Kirchoff's Current Law.

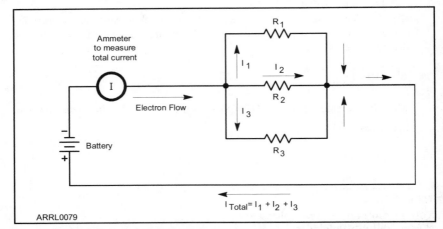

ARRL0079

**Figure G5-1** — The sum of the current flowing into a junction point (node) in a circuit must be equal to the current flowing out of the junction point (node). This principle is called Kirchhoff's Current Law, named for Gustav Kirchhoff, the German scientist who discovered it.

**G5B03** How many watts of electrical power are used if 400 VDC is supplied to an 800-ohm load?

A. 0.5 watts
B. 200 watts
C. 400 watts
D. 3200 watts

**(B)** Since P = I × E and I = E/R, the power in a circuit also can be expressed as

$$P = \frac{E \times E}{R}$$

In this case:

$$P = \frac{E \times E}{R} = \frac{400 \times 400}{800} = \frac{160,000}{800} = 200 \text{ W}$$

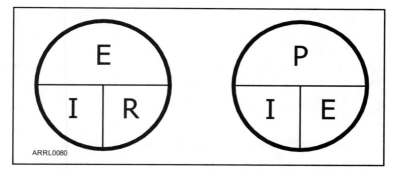

ARRL0080

**Figure G5-2 — The "Ohm's Law Circle" and the "Power Circle" will help you remember the equations that include voltage, current, resistance and power. Cover the letter representing the unknown quantity to find an equation to calculate that quantity. If you cover the I in the Ohm's Law Circle, you are left with E/R. If you cover the P in the Power Circle, you are left with I × E. Combining these terms, you can write the equation to calculate power when you know the voltage and the resistance.**

**G5B04** How many watts of electrical power are used by a 12-VDC light bulb that draws 0.2 amperes?

A. 2.4 watts
B. 24 watts
C. 6 watts
D. 60 watts

**(A)** Use the Power Circle drawing to find the equation to calculate power. The power in a circuit is equal to the voltage times the current: P = I × E = 0.2 × 12 = 2.4 watts

**G5B05**  How many watts are dissipated when a current of
7.0 milliamperes flows through 1.25 kilohms?

A.  Approximately 61 milliwatts
B.  Approximately 61 watts
C.  Approximately 11 milliwatts
D.  Approximately 11 watts

**(A)** Use the Ohm's Law Circle and Power Circle drawings to find the equations to calculate power. Since P = I × E and E = R × I, the power in a circuit can also be expressed as P = I × I × R, so P = 0.007 × 0.007 × 1250 = 0.06125 W = 61.25 mW. Remember that 7 milliamperes is equal to 0.007 ampere, 1.25 kilohms is equal to 1250 ohms and 0.06125 watt is equal to approximately 61 milliwatts.

**G5B06**  What is the output PEP from a transmitter if an oscilloscope
measures 200 volts peak-to-peak across a 50-ohm dummy load
connected to the transmitter output?

A.  1.4 watts
B.  100 watts
C.  353.5 watts
D.  400 watts

**(B)** PEP is the output power measured at the peak of the RF cycle at the peak of the signal's envelope and is equal to:

$$PEP = \frac{(\text{Peak envelope voltage} \times 0.707)^2}{R}$$

Peak envelope voltage = Peak-to-peak envelope voltage / 2, so:

$$PEP = \frac{(100 \times 0.707)^2}{50} = 100 \text{ W}$$

**G5B07**  Which value of an AC signal results in the same power
dissipation as a DC voltage of the same value?

A.  The peak-to-peak value
B.  The peak value
C.  The RMS value
D.  The reciprocal of the RMS value

**(C)** RMS, or root mean square, voltage values convert a constantly-varying ac voltage to the equivalent of a constant dc voltage. The RMS value of an ac voltage is the value that would deliver the same amount of power to a resistance as a dc voltage of the same value.

**G5B08**   What is the peak-to-peak voltage of a sine wave that has an RMS voltage of 120 volts?

A. 84.8 volts
B. 169.7 volts
C. 240.0 volts
D. 339.4 volts

**(D)** If you know the RMS voltage and want to know the peak value, multiply the RMS by the square root of 2 (which is 1.414). If you want to know peak-to-peak, double the result. In this case, 120 × 1.414 × 2 = 339.4 V.

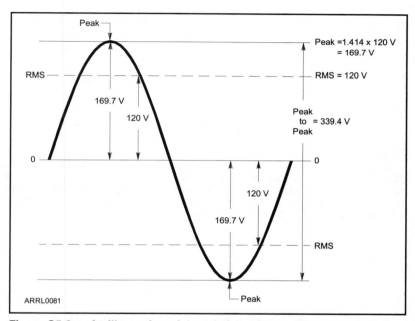

Figure G5-3 — An illustration of the relationship between ac measurements for sine wave waveforms. RMS value conversions assume that the waveform is a sine wave. Different conversion values are required for other waveform shapes, such as square or triangle waves.

**G5B09** What is the RMS voltage of a sine wave with a value of 17 volts peak?

A. 8.5 volts
B. 12 volts
C. 24 volts
D. 34 volts

**(B)** If you know the peak voltage, you can find the RMS value by multiplying the peak voltage by 0.707 (which is the same as dividing by the square root of 2).

$17 \times 0.707 = 12$ V

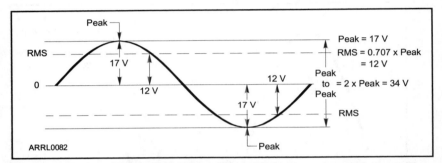

**Figure G5-4 — An illustration of the relationship between ac measurements for sine wave waveforms. RMS value conversions assume that the waveform is a sine wave. Different conversion values are required for other waveform shapes, such as square or triangle waves.**

**G5B10** **What percentage of power loss would result from a transmission line loss of 1 dB?**

A. 10.9%
B. 12.2%
C. 20.5%
D. 25.9%

**(C)** The decibel scale is a logarithmic scale in which a two-times increase (or decrease) in power is represented by 3 dB. The mathematical formula for the decibel scale for power is:

$$dB = 10 \times \log_{10} \left( \frac{P_2}{P_1} \right)$$

where $P_1$ = reference power and $P_2$ = power being compared to the reference value. In this question we are told that the transmission line to an antenna results in a signal loss of 1 dB. We can use the decibel equation to calculate the power that reaches the antenna through this feed line, and from that we can calculate the power lost in the feed line. If we assume a transmitter power of 100 W, then the resulting answer will be the percentage of the original power that is lost in the feed line. First we will solve the decibel equation for $P_2$, the power that reaches the antenna.

$$dB = 10 \times \log_{10} \left( \frac{P_2}{P_1} \right)$$

$$\frac{dB}{10} = \log_{10} \left( \frac{P_2}{P_1} \right)$$

$$\log_{10}^{-1} \left( \frac{dB}{10} \right) = \left( \frac{P_2}{P_1} \right)$$

Note: The notation that shows the logarithm raised to the negative 1 power means the antilog, or inverse logarithm. On some scientific calculators this button is also labeled $10^x$, which means "raise 10 to the power of this value." The decibel value is given as a loss of 1 dB, so we will write that as −1 dB.

$$P_1 \ \log_{10}^{-1} \left( \frac{dB}{10} \right) = P_2 = 100 \ W \times \log_{10}^{-1} \left( \frac{-1 \ dB}{10} \right)$$

$$= 100 \ W \times 0.794 = 79.4 \ W$$

If 79.4 W is the power actually reaching the antenna, then we can calculate the lost power by subtracting this value from the original power: 100 W − 79.4 W = 20.6 W. Because we used 100 W as the reference power, this value is the percentage of the original power that is lost in our feed line, when it has a loss of 1 dB. The percentage of the power lost in this feed line is 20.6%.

**G5B11**  What is the ratio of peak envelope power to average power for an unmodulated carrier?

A. .707
B. 1.00
C. 1.414
D. 2.00

**(B)** It's 1.0 because for an unmodulated carrier all RF cycles have the same voltage, meaning that the envelope's average and peak values are the same.

**G5B12**  What would be the RMS voltage across a 50-ohm dummy load dissipating 1200 watts?

A. 173 volts
B. 245 volts
C. 346 volts
D. 692 volts

**(B)** It's 245 V because $P = E^2 / R$, so $E = \sqrt{1200 \times 50}$. This is the RMS voltage across the 50-ohm load.

**G5B13**  What is the output PEP of an unmodulated carrier if an average reading wattmeter connected to the transmitter output indicates 1060 watts?

A. 530 watts
B. 1060 watts
C. 1500 watts
D. 2120 watts

**(B)** The PEP and average power of an unmodulated carrier are the same. For an unmodulated carrier all RF cycles have the same voltage, meaning that the envelope's average and peak values are the same.

**G5B14**  What is the output PEP from a transmitter if an oscilloscope measures 500 volts peak-to-peak across a 50-ohm resistor connected to the transmitter output?

A. 8.75 watts
B. 625 watts
C. 2500 watts
D. 5000 watts

**(B)** $PEP = (E_{RMS})^2 / R = (250 \times 0.707)^2 / 50 = 625$ W

## G5C Resistors, capacitors and inductors in series and parallel; transformers

**G5C01** What causes a voltage to appear across the secondary winding of a transformer when an AC voltage source is connected across its primary winding?

A. Capacitive coupling
B. Displacement current coupling
C. Mutual inductance
D. Mutual capacitance

**(C)** A transformer consists of two coils (windings) sharing a common core so that the flux from one winding is shared by both windings. An ac signal in the input coil, called the primary winding produces an ac signal in the output coil, called the secondary winding. The core material might be layers of steel, a powdered iron mixture, some other magnetic material, or even air. The coupling between the primary and secondary windings is called mutual inductance.

When a current flows through the primary winding it creates a magnetic field in the core. That magnetic field changes polarity and strength as the primary ac voltage changes. The changing magnetic field in the common core is shared by the secondary winding, inducing a voltage across the turns of the secondary winding and creating a current in the secondary circuit.

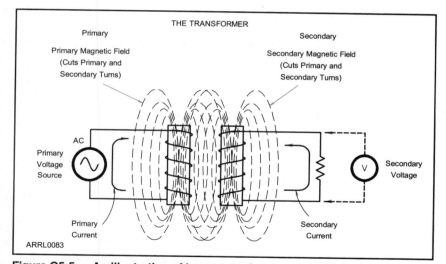

Figure G5-5 — An illustration of how a transformer works. The input of a transformer is called the primary winding, and the output is called the secondary winding. In this drawing separate primary and secondary cores are shown to illustrate how the windings share magnetic flux. In most transformers, both windings are wound on a common core for more complete sharing of flux.

**G5C02**  Which part of a transformer is normally connected to the incoming source of energy?

A. The secondary
B. The primary
C. The core
D. The plates

**(B)** The primary winding is generally considered to be the input. However energy can flow in either direction, primary-to-secondary or secondary-to-primary.

**G5C03**  Which of the following components should be added to an existing resistor to increase the resistance?

A. A resistor in parallel
B. A resistor in series
C. A capacitor in series
D. A capacitor in parallel

**(B)** Add additional resistors in series to increase the total resistance.

**G5C04**  What is the total resistance of three 100-ohm resistors in parallel?

A. .30 ohms
B. .33 ohms
C. 33.3 ohms
D. 300 ohms

**(C)** To calculate the total resistance of resistors in parallel take the reciprocal of the sum of the resistor's reciprocal values as follows:

$$R = \frac{1}{\dfrac{1}{100} + \dfrac{1}{100} + \dfrac{1}{100}} = \frac{100}{3} = 33.3 \ \Omega$$

**G5C05**  If three equal value resistors in parallel produce 50 ohms of resistance, and the same three resistors in series produce 450 ohms, what is the value of each resistor?

A. 1500 ohms
B. 90 ohms
C. 150 ohms
D. 175 ohms

**(C)** For N resistors of equal value in parallel the resulting equivalent resistance equals the common resistor value divided by the number of resistors:

R = Common value / N.

For N resistors of equal value in series the resulting equivalent resistance equals the sum of the resistor values:

R = R1 + R2 + R3 +... = Common value × N

In this case, in parallel the total resistance of 50 Ω = Common Value / 3. In series, the total resistance of 450 Ω = Common Value × 3. The common value can be calculated from either equation:

Common value = 3 × 50 = 150 Ω or Common value = 450 / 3 = 150 Ω.

**G5C06**  What is the RMS voltage across a 500-turn secondary winding in a transformer if the 2250-turn primary is connected to 120 VAC?

A. 2370 volts
B. 540 volts
C. 26.7 volts
D. 5.9 volts

**(C)** The voltage in the secondary winding of a transformer is equal to the voltage in the primary winding times the ratio of windings in the secondary to the primary.

$$E_S = E_P \times \frac{N_S}{N_P}$$

If the 2250-turn primary is connected to 120 volts ac, the voltage across a 500-turn secondary winding in the transformer is 26.7 volts:

$$E_S = E_P \times \frac{N_S}{N_P} = 120 \text{ V} \times \frac{500}{2250} = 26.7 \text{ V}$$

**G5C07**  What is the turns ratio of a transformer used to match an audio amplifier having a 600-ohm output impedance to a speaker having a 4-ohm impedance?

   A.  12.2 to 1
   B.  24.4 to 1
   C.  150 to 1
   D.  300 to 1

**(A)**  To calculate the turns ratio:

$$\text{Turns ratio} = \sqrt{\frac{N_P}{N_S}} = \sqrt{\frac{600}{4}} = \sqrt{150} = 12.2$$

**G5C08**  What is the equivalent capacitance of two 5000 picofarad capacitors and one 750 picofarad capacitor connected in parallel?

   A.  576.9 picofarads
   B.  1733 picofarads
   C.  3583 picofarads
   D.  10750 picofarads

**(D)**  To calculate the total capacitance of capacitors in parallel add the values of the capacitors together,

$C = C1 + C2 + C3 + \ldots$

In this case:

$C = 5000 \text{ pF} + 5000 \text{ pF} + 750 \text{ pF} = 10750 \text{ pF}$

**G5C09**  What is the capacitance of three 100 microfarad capacitors connected in series?

   A.  .30 microfarads
   B.  .33 microfarads
   C.  33.3 microfarads
   D.  300 microfarads

**(C)**  For N capacitors of equal value in series, the resulting capacitance equals the common capacitor value divided by the number of capacitors,

$C = \text{Common value} / N$

$C = 100 \ \mu\text{F} / 3 = 33.3 \ \mu\text{F}$

**G5C10** What is the inductance of three 10 millihenry inductors connected in parallel?

A. .30 Henrys
B. 3.3 Henrys
C. 3.3 millihenrys
D. 30 millihenrys

(C) For N inductors of equal value in parallel, the resulting inductance equals the common inductor value divided by the number of inductors,

L = Common value / N

L = 10 mH / 3 = 3.3 mH

**G5C11** What is the inductance of a 20 millihenry inductor in series with a 50 millihenry inductor?

A. .07 millihenrys
B. 14.3 millihenrys
C. 70 millihenrys
D. 1000 millihenrys

(C) For N inductors in series, the resulting inductance equals the sum of the inductor values,

L = L1 + L2 + L3 +... = Common value × N

L = 20 mH + 50 mH = 70 mH

**G5C12** What is the capacitance of a 20 microfarad capacitor in series with a 50 microfarad capacitor?

A. .07 microfarads
B. 14.3 microfarads
C. 70 microfarads
D. 1000 microfarads

(B) To calculate the total capacitance of capacitors in series take the reciprocal of the sum of the capacitor's reciprocal values as follows:

$$C = \cfrac{1}{\dfrac{1}{C_1} + \dfrac{1}{C_2} + \cdots + \dfrac{1}{C_N}}$$

In the case of two capacitors in series, the equation simplifies to:

$$C = \frac{C_1 \times C_2}{C_1 + C_2} = \frac{20 \times 50}{20 + 50} = \frac{1000}{70} = 14.3 \; \mu F$$

**G5C13** Which of the following components should be added to a capacitor to increase the capacitance?

A. An inductor in series
B. A resistor in series
C. A capacitor in parallel
D. A capacitor in series

**(C)** Add additional capacitors in parallel to increase the total capacitance.

**G5C14** Which of the following components should be added to an inductor to increase the inductance?

A. A capacitor in series
B. A resistor in parallel
C. An inductor in parallel
D. An inductor in series

**(D)** Add additional inductors in series to increase the total inductance.

**G5C15** What is the total resistance of a 10 ohm, a 20 ohm, and a 50 ohm resistor in parallel?

A. 5.9 ohms
B. 0.17 ohms
C. 10000 ohms
D. 80 ohms

**(A)** To calculate the total resistance of resistors in parallel take the reciprocal of the sum of the resistor's reciprocal values as follows:

$$R = \frac{1}{\frac{1}{10} + \frac{1}{20} + \frac{1}{50}} = 5.9 \ \Omega$$

# Circuit Components

Your General class exam (Element 3) will consist of 35 questions taken from the General class question pool as prepared by the Volunteer Examiner Coordinators' Questions Pool Committee. A certain number of questions are taken from each of the 10 subelements. There will be 3 questions from the subelement shown in this chapter. These questions are divided into 3 groups, labeled G6A through G6C.

## SUBELEMENT G6 — CIRCUIT COMPONENTS
### [3 Exam Questions — 3 Groups]

### G6A    Resistors; capacitors; inductors

**G6A01**  Which of the following is an important characteristic for capacitors used to filter the DC output of a switching power supply?

A. Low equivalent series resistance
B. High equivalent series resistance
C. Low Temperature coefficient
D. High Temperature coefficient

**(A)** It is important to use filter capacitors that have a low equivalent series resistance (ESR) rating for the output filter of a switching power supply. Low ESR means that currents flowing into and out of the capacitor as it smoothes the output voltage will not cause ripple (voltage variations) and losses that heat the capacitor.

**G6A02**  Which of the following types of capacitors are often used in power supply circuits to filter the rectified AC?

A. Disc ceramic
B. Vacuum variable
C. Mica
D. Electrolytic

**(D)** Because electrolytic capacitors can be made with the required high capacitance value in a small package, they are often used in power supply circuits to filter the rectified ac voltage.

**G6A03** Which of the following is an advantage of ceramic capacitors as compared to other types of capacitors?

A. Tight tolerance
B. High stability
C. High capacitance for given volume
D. Comparatively low cost

**(D)** The primary advantage of ceramic capacitors is that they offer good performance at low cost. Ceramic capacitors are usually used as bypass capacitors that filter out high-frequency ac voltages from dc and low-frequency signal connections.

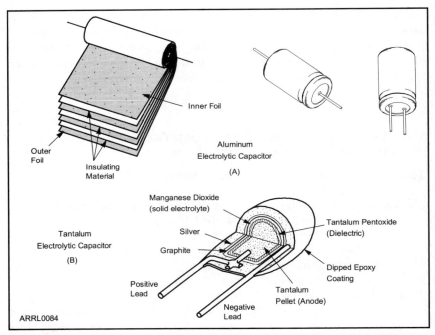

**Figure G6-1 — Part A shows the construction of an aluminum electrolytic capacitor. Part B shows the construction of a tantalum electrolytic capacitor.**

**G6A04** Which of the following is an advantage of an electrolytic capacitor?

A. Tight tolerance
B. Non-polarized
C. High capacitance for given volume
D. Inexpensive RF capacitor

**(C)** Electrolytic capacitors are designed to provide large values of capacitance for energy storage and ac voltage filtering. Their construction, shown in the illustration, provides relatively high capacitance in a small volume. The tradeoff is that dc voltage applied to them must always be of the same polarity and the manufacturing technique leads to rather wide variations in capacitance.

**G6A05** Which of the following is one effect of lead inductance in a capacitor used at VHF and above?

A. Effective capacitance may be reduced
B. Voltage rating may be reduced
C. ESR may be reduced
D. The polarity of the capacitor might become reversed

**(A)** Lead inductance creates a small inductance in series with the capacitor. At low frequencies, lead inductance can be ignored but at VHF and higher frequencies, the inductive reactance begins to cancel the capacitor's own reactance, reducing the effective capacitance dramatically! Capacitors to be used at these high frequencies must have very short leads.

**G6A06** What will happen to the resistance if the temperature of a resistor is increased?

A. It will change depending on the resistor's reactance coefficient
B. It will stay the same
C. It will change depending on the resistor's temperature coefficient
D. It will become time dependent

**(C)** The resistance of almost any substance changes when heated. The amount of change in a resistor's value depends on its temperature coefficient. Most resistors have a positive temperature coefficient, meaning that resistance increases with temperature.

**G6A07** Which of the following is a reason not to use wire-wound resistors in an RF circuit?

A. The resistor's tolerance value would not be adequate for such a circuit

B. The resistor's inductance could make circuit performance unpredictable

C. The resistor could overheat

D. The resistor's internal capacitance would detune the circuit

**(B)** Wire-wound resistors are made exactly as their name implies. A length of "resistance wire" — usually Nichrome — is wound around a ceramic form. The wire has some specific resistance per length of wire, so the length of wire is chosen to give the desired resistance. Because a long length of wire is difficult to accommodate in an electronics circuit, the wire is wound on the form to reduce the overall length of the resistor. The resistor is then coated with a ceramic or other insulating material to protect the wire. If this construction method sounds like an inductor to you, you are absolutely correct! It is not a good idea to use wire-wound resistors in any RF circuits or anywhere you don't want some extra amount of inductance to be included in the circuit. The extra inductance will detune any resonant circuit of which it is part or add unwanted inductive reactance to signal paths.

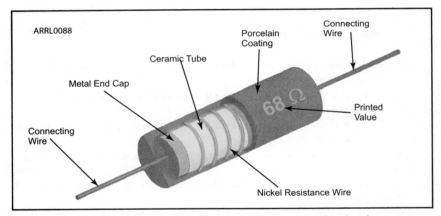

**Figure G6-2 — This drawing shows the construction of a simple wire-wound resistor.**

**G6A08**  Which of the following describes a thermistor?

A.  A resistor that is resistant to changes in value with temperature variations

B.  A device having a specific change in resistance with temperature variations

C.  A special type of transistor for use at very cold temperatures

D.  A capacitor that changes value with temperature

**(B)**  All resistors change value to some degree with temperature. A thermistor is a special type of resistor whose change in value with temperature is precisely controlled so that it can be used as a temperature sensor.

**G6A09**  What is an advantage of using a ferrite core toroidal inductor?

A.  Large values of inductance may be obtained

B.  The magnetic properties of the core may be optimized for a specific range of frequencies

C.  Most of the magnetic field is contained in the core

D.  All of these choices are correct

**(D)**  Because a toroid core offers a continuous, circular path for magnetic flux, nearly all of a toroidal inductor's magnetic field in contained inside the core. This gives a toroidal inductor a self-shielding property that makes them ideal for use in RF circuits, where you do not want interaction between nearby inductors.

Ferrite is a ceramic containing iron compounds. Ferrite and powdered-iron toroid cores both have high permeabilities (the ability to store magnetic energy) make it possible to obtain large values of inductance in a relatively small package, as compared to the number of turns that would be required with an air-core inductor. Toroid cores are made in different "mixes" for use over various frequency ranges.

All of the answer choices are true statements, so answer D is correct.

**G6A10** How should the winding axes of solenoid inductors be placed to minimize their mutual inductance?

A. In line
B. Parallel to each other
C. At right angles
D. Interleaved

**(C)** An inductor that is formed by winding a coil of wire on a straight coil form of either air or a magnetic material is called a solenoid inductor. The magnetic field associated with a solenoid core extends through the center of the core and then around the outside of the coil. To minimize any interaction between the magnetic fields around solenoid inductors in a circuit, the coils should be placed so that the axes along which the wire is wound are at right angles to each other. (To ensure minimum interaction between solenoid coils you might also have to place each inductor inside its own shielded enclosure.)

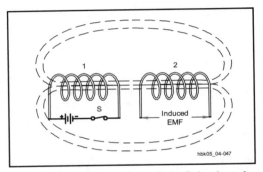

Figure G6-3 — When the switch, S, is closed, current flows through coil number 1, setting up a shared magnetic field that causes a voltage to be included in the runs of coil number 2. Mutual inductance is created by the shared magnetic field.

**G6A11** Why would it be important to minimize the mutual inductance between two inductors?

A. To increase the energy transfer between circuits
B. To reduce unwanted coupling between circuits
C. To reduce conducted emissions
D. To increase the self-resonant frequency of the inductors

**(B)** It is important to minimize unwanted mutual inductance between two inductors to prevent coupling — transferring energy — between different circuit stages.

**G6A12**   What is a common name for an inductor used to help smooth the DC output from the rectifier in a conventional power supply?

A. Back EMF choke
B. Repulsion coil
C. Charging inductor
D. Filter choke

**(D)** A filter choke is an inductor that is connected in series with the output of a power supply. It works with the filter capacitor to smooth the rectified ac from the rectifier into dc. The inductor is called a choke because it "chokes off" the variations in voltage at its input.

**G6A13**   What is an effect of inter-turn capacitance in an inductor?

A. The magnetic field may become inverted
B. The inductor may become self resonant at some frequencies
C. The permeability will increase
D. The voltage rating may be exceeded

**(B)** Inter-turn capacitance is created by adjacent turns of wire separated by air or their insulation. Even though the turns of wire are connected, the small separation between the wire surfaces creates a small capacitance. In a coil of many turns, the inter-turn capacitance can become significant. At high frequencies, the combination of the coil's inductance and the inter-turn capacitance can become a series- or parallel-resonant circuit!

**G6B   Rectifiers; solid state diodes and transistors; vacuum tubes; batteries**

**G6B01**   What is the peak-inverse-voltage rating of a rectifier?

A. The maximum voltage the rectifier will handle in the conducting direction
B. 1.4 times the AC frequency
C. The maximum voltage the rectifier will handle in the non-conducting direction
D. 2.8 times the AC frequency

**(C)** Power-supply rectifier diodes are made from semiconductor material. Diodes allow current to flow only in one direction. If the voltage is high enough, however, current can be forced to flow in the opposite direction, often destroying the diode. The peak inverse voltage rating (PIV) of a rectifier diode is the maximum reverse voltage that should ever be applied to the diode — it is guaranteed to withstand that voltage.

**G6B02**  What are two major ratings that must not be exceeded for silicon diode rectifiers?

A.  Peak inverse voltage; average forward current
B.  Average power; average voltage
C.  Capacitive reactance; avalanche voltage
D.  Peak load impedance; peak voltage

(A) Peak inverse voltage and average forward current are two major ratings that must not be exceeded for silicon diode rectifiers. (Peak inverse voltage rating (PIV) is described in question G6B01.) The maximum average forward current is the other major rating that must not be exceeded. Forward current causes power to be dissipated in the diode due to the forward voltage drop (0.7 volts in silicon rectifiers). The amount of power is 0.7 V times the average forward current. Too much current overheats and destroys the diode.

**G6B03**  What is the approximate junction threshold voltage of a germanium diode?

A.  0.1 volts
B.  0.3 volts
C.  0.7 volts
D.  1.0 volts

(B) The junction threshold voltage is the voltage at which a diode begins to conduct significant current across its PN junction. The amount of voltage depends on the material from which the diode is constructed. The junction threshold voltage of silicon diodes is approximately 0.7 V and for germanium diodes approximately 0.3 V.

**G6B04** When two or more diodes are connected in parallel to increase current handling capacity, what is the purpose of the resistor connected in series with each diode?

A. To ensure the thermal stability of the power supply
B. To regulate the power supply output voltage
C. To ensure that one diode doesn't carry most of the current
D. To act as an inductor

**(C)** Two or more diodes are sometimes connected in parallel to increase the current-handling ability of a circuit. There should always be a small resistor connected in series with each parallel diode. Without a resistor in series with each diode, one diode may conduct most of the current because of small variations in threshold voltage and that could destroy the diode by exceeding its average forward current rating. The resistor value should be chosen to provide a few tenths of a volt drop at the expected forward current.

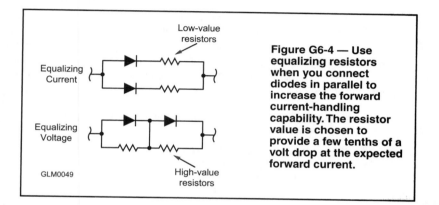

Figure G6-4 — Use equalizing resistors when you connect diodes in parallel to increase the forward current-handling capability. The resistor value is chosen to provide a few tenths of a volt drop at the expected forward current.

**G6B05** What is the approximate junction threshold voltage of a conventional silicon diode?

A. 0.1 volts
B. 0.3 volts
C. 0.7 volts
D. 1.0 volts

**(C)** See the discussion for question G6B03.

**G6B06** Which of the following is an advantage of using a Schottky diode in an RF switching circuit as compared to a standard silicon diode?

A. Lower capacitance
B. Lower inductance
C. Longer switching times
D. Higher breakdown voltage

**(A)** The construction of a Schottky diode results in much lower capacitance between the cathode and the anode. This makes the diode respond better to high-frequency signals in a switching power supply, digital logic, or an RF circuit.

**G6B07** What are the stable operating points for a bipolar transistor used as a switch in a logic circuit?

A. Its saturation and cut-off regions
B. Its active region (between the cut-off and saturation regions)
C. Its peak and valley current points
D. Its enhancement and deletion modes

**(A)** When a bipolar transistor is used as a switch in a logic circuit it is important that the switch either be "all the way on" or "all the way off." This is achieved by operating the transistor either in its saturation region to turn the switch on, or in its cut-off region to turn the switch off.

**G6B08** Why must the cases of some large power transistors be insulated from ground?

A. To increase the beta of the transistor
B. To improve the power dissipation capability
C. To reduce stray capacitance
D. To avoid shorting the collector or drain voltage to ground

**(D)** On some power transistors the collector is connected to the case to improve the transfer of heat away from the semiconductor material. This helps keep the transistor from overheating when controlling high currents but requires that the case of the transistor (usually connected to the collector) be insulated from ground since most circuits do not ground the collector.

**G6B09** Which of the following describes the construction of a MOSFET?

A. The gate is formed by a back-biased junction
B. The gate is separated from the channel with a thin insulating layer
C. The source is separated from the drain by a thin insulating layer
D. The source is formed by depositing metal on silicon

**(B)** A MOSFET (Metal Oxide Semiconductor Field Effect Transistor) is similar to a JFET (Junction Field Effect Transistor), but instead of the gate electrode being in direct contact with the channel between drain and source, it is insulated by a thin insulating layer of oxide. The gate voltage still controls electron flow between the drain and source, but very little current flows in the gate circuit.

**G6B10** Which element of a triode vacuum tube is used to regulate the flow of electrons between cathode and plate?

A. Control grid
B. Heater
C. Screen Grid
D. Trigger electrode

**(A)** The control grid is closest to the cathode, the element of the tube that generates the electrons. By varying the control grid voltage with respect to the cathode voltage, electron flow between cathode and plate can be controlled.

**G6B11** Which of the following solid state devices is most like a vacuum tube in its general operating characteristics?

A. A bipolar transistor
B. A Field Effect Transistor
C. A tunnel diode
D. A varistor

**(B)** A Field-Effect Transistor (FET) acts very much like a vacuum tube with the gate electrode taking the role of the control grid, controlling electron flow between the drain and source that correspond to the plate and cathode, respectively.

**G6B12** What is the primary purpose of a screen grid in a vacuum tube?

A. To reduce grid-to-plate capacitance
B. To increase efficiency
C. To increase the control grid resistance
D. To decrease plate resistance

**(A)** The screen grid is placed between the plate and control grid and kept at a constant voltage. This isolates the control grid from the plate and reduces the capacitance between them.

**G6B13** What is an advantage of the low internal resistance of nickel-cadmium batteries?

A. Long life
B. High discharge current
C. High voltage
D. Rapid recharge

**(B)** Nickel-cadmium (NiCd) batteries are constructed so that they can supply large quantities of current very quickly. This makes them useful in portable power tools and radio transceivers.

**G6B14**  What is the minimum allowable discharge voltage for maximum life of a standard 12 volt lead acid battery?

A. 6 volts
B. 8.5 volts
C. 10.5 volts
D. 12 volts

**(C)** Standard 12-volt lead-acid batteries are composed of six 2-volt cells connected in series. Each cell should not be discharged below 1.75 volts to avoid causing irreversible chemical changes that damage the cell. Thus, the minimum voltage for a standard 12-volt battery is 6 × 1.75 = 10.5 volts.

**G6B15**  When is it acceptable to recharge a carbon-zinc primary cell?

A. As long as the voltage has not been allowed to drop below 1.0 volt
B. When the cell is kept warm during the recharging period
C. When a constant current charger is used
D. Never

**(D)** Carbon-zinc batteries are not rechargeable because the chemical reaction that powers them cannot be reversed. Do not attempt to recharge them! If current is forced back through the cell, the resulting chemical reaction releases acid and heat that may breach the battery's case, damaging whatever houses the battery.

## G6C  Analog and digital integrated circuits (IC's); microprocessors; memory; I/O devices; microwave IC's (MMIC's); display devices

**G6C01**  Which of the following is an analog integrated circuit?

A. NAND Gate
B. Microprocessor
C. Frequency Counter
D. Linear voltage regulator

**(D)** A linear voltage regulator is composed of a transistor, a voltage reference diode, and an amplifier that compares the circuit output voltage to the voltage reference. The regulator continuously varies the transistor current to keep the output voltage constant. Operation over a continuous range of voltages and currents is what analog integrated circuits do.

**G6C02** **What is meant by the term MMIC?**

A. Multi Megabyte Integrated Circuit
B. Monolithic Microwave Integrated Circuit
C. Military-specification Manufactured Integrated Circuit
D. Mode Modulated Integrated Circuit

**(B)** An MMIC is a special type of analog IC containing circuits to perform RF operations such as amplification, modulation and demodulation, and mixing at HF through microwave frequencies. Some MMICs combine several functions, acting as an entire receiver front end, for example. The MMIC is what enables communications engineers to construct low-cost, hand-held mobile phones and GPS receivers, among other sophisticated examples of wireless technology.

**G6C03** **Which of the following is an advantage of CMOS integrated circuits compared to TTL integrated circuits?**

A. Low power consumption
B. High power handling capability
C. Better suited for RF amplification
D. Better suited for power supply regulation

**(A)** Integrated circuits based on CMOS (Complementary Metal Oxide Semiconductor) technology consume very little current when operating, only drawing significant power when switching between ON and OFF states.

**G6C04** **What is meant by the term ROM?**

A. Resistor Operated Memory
B. Read Only Memory
C. Random Operational Memory
D. Resistant to Overload Memory

**(B)** ROM, or read only memory, contains data or program instructions that do not need to be changed. ROM is used to prevent changes.

**G6C05** **What is meant when memory is characterized as "non-volatile"?**

A. It is resistant to radiation damage
B. It is resistant to high temperatures
C. The stored information is maintained even if power is removed
D. The stored information cannot be changed once written

**(C)** Non-volatile memory can retain stored data even if power is removed and restored. It never needs to have the data "refreshed." Volatile memory loses the data if power is removed and often needs to have the data refreshed while in operation, such as for dynamic types of memory.

**G6C06**   Which of the following describes an integrated circuit operational amplifier?

A. Digital
B. MMIC
C. Programmable Logic
D. Analog

**(D)** An operational amplifier, or op amp, is an analog circuit, operating over a continuous range of voltage and current.

**G6C07**   What is one disadvantage of an incandescent indicator compared to an LED?

A. Low power consumption
B. High speed
C. Long life
D. High power consumption

**(D)** Incandescent indicators (lamps) are bright but they waste as heat most of the energy supplied to them. For the same amount of light output, LEDs are much more power-efficient than incandescents.

**G6C08**   How is an LED biased when emitting light?

A. Beyond cutoff
B. At the Zener voltage
C. Reverse Biased
D. Forward Biased

**(D)** An LED emits light when forward biased so that current flows through the PN-junction of the diode. Photons of light are given off when the electrons from the N-type material combine with the holes in the P-type material.

**G6C09**   Which of the following is a characteristic of a liquid crystal display?

A. It requires ambient or back lighting
B. It offers a wide dynamic range
C. It has a wide viewing angle
D. All of these choices are correct

**(A)** A liquid crystal display (LCD) works by blocking the transmission of light through an otherwise transparent layer of liquid crystals. Transparent electrodes are printed on the glass layers on either side of the liquid crystals to form the pattern of the digits, characters, and symbols. When voltage of the right polarity is applied between the electrodes, the liquid crystals twist into a pattern that blocks light. This is why an LCD requires ambient light to reflect off the back of the display or an active source of light behind the liquid crystals (backlighting) in order to see the desired pattern.

**G6C10** **What two devices in an Amateur Radio station might be connected using a USB interface?**

A. Computer and transceiver
B. Microphone and transceiver
C. Amplifier and antenna
D. Power supply and amplifier

**(A)** A USB (Universal Serial Bus) interface is a means of exchanging digital data in a serial stream (meaning one bit at a time). The computer and transceiver are the most likely to communicate over such an interface.

**G6C11** **What is a microprocessor?**

A. A low power analog signal processor used as a microwave detector
B. A computer on a single integrated circuit
C. A microwave detector, amplifier, and local oscillator on a single integrated circuit
D. A low voltage amplifier used in a microwave transmitter modulator stage

**(B)** These miniature computers on a single IC consist of thousands of digital gates and logic elements that can retrieve program instructions (known as machine language) and execute them using digital data. Simple microprocessors may be little more than a set of registers that store and exchange data for the program instructions. Complex microprocessors may contain sub-processors that are dedicated to special functions, sophisticated communications interfaces, or include analog-digital converters to provide an all-in-one DSP platform.

**G6C12** **Which of the following connectors would be a good choice for a serial data port?**

A. PL-259
B. Type N
C. Type SMA
D. DE-9

**(D)** RS-232 serial data interfaces are common in amateur equipment. The most common connector used for these interfaces is the 9-pin D-style DE-9, also called a DB-9. These connectors are usually found on personal computers as COM ports. RS-232 interfaces are being phased out in favor of USB and other interfaces that can transfer data at higher rates.

**G6C13**  Which of these connector types is commonly used for RF service at frequencies up to 150 MHz?

A. Octal
B. RJ-11
C. PL-259
D. DB-25

(**C**) The UHF connector family includes the PL-259 cable plug and SO-239 chassis-mounted socket. It is the most popular type of RF connector used on amateur equipment. UHF does not refer to a frequency range in this case.

Figure G6-5 — This photo shows some common coaxial-cable connectors. At left is a BNC connector pair. Many hand-held radios use BNC connectors. They are a popular connector for RG-58-size cable. In the center is a pair of type N connectors. These are often used for UHF equipment because of their low loss. Type N connectors provide a weatherproof connector for RG-8-size cables. At the right is a PL-259 coaxial connector and its mating SO-239 chassis connector. Most HF equipment uses these connectors. They are designed for use with RG-8-size cables, although reducer adapters are available for smaller-diameter cables such as RG-58 and RG-59. Although the PL-259 is called a "UHF connector," it is seldom used on UHF equipment because it is considered to have too much loss for UHF. UHF in this case does not refer to a frequency range.

**G6C14**  Which of these connector types is commonly used for audio signals in Amateur Radio stations?

A. PL-259
B. BNC
C. RCA Phono
D. Type N

(**C**) The RCA phono connector is the most common audio signal connector for consumer electronics and a great deal of amateur equipment. The connector's name derives from its early use by the RCA Company for audio connectors and its subsequent popularity for the connection of phonographs to amplifiers and receivers.

**G6C15**   What is the main reason to use keyed connectors instead of non-keyed types?

A. Prevention of use by unauthorized persons
B. Reduced chance of incorrect mating
C. Higher current carrying capacity
D. All of these choices are correct

(B) A keyed connector has an asymmetrical body shape or arrangement of contacts. This prevents reversed or improper circuit connections that lead to improper operation or damage to the equipment. Do not force keyed connectors together if they are not oriented correctly.

**G6C16**   Which of the following describes a type-N connector?

A. A moisture-resistant RF connector useful to 10 GHz
B. A small bayonet connector used for data circuits
C. A threaded connector used for hydraulic systems
D. An audio connector used in surround-sound installations

(A) Type N connectors are connectors for coaxial cable used at HF, VHF, UHF and microwave frequencies up to 10 GHz. The type N connector shell is threaded like a PL-259, but the special design of its body presents the same 50-$\Omega$ impedance as coaxial cable so no signals are reflected or lost at the junction of connector and feed line. Type N connectors also have special gaskets built-in so that they are waterproof without requiring additional coatings.

**G6C17**   What is the general description of a DIN type connector?

A. A special connector for microwave interfacing
B. A DC power connector rated for currents between 30 and 50 amperes
C. A family of multiple circuit connectors suitable for audio and control signals
D. A special watertight connector for use in marine applications

(C) Multiple-pin DIN and Mini-DIN connectors are the standard for accessory connectors on amateur equipment. DIN stands for the Deutsches Institut für Normung, the German standards organization. DIN and Mini-DIN connectors are keyed and have up to 9 pins.

**G6C18**   What is a type SMA connector?

A. A large bayonet-type connector usable at power levels in excess of 1 kW
B. A small threaded connector suitable for signals up to several GHz
C. A connector designed for serial multiple access signals
D. A type of push-on connector intended for high-voltage applications

(B) SMA connectors are small threaded connectors designed for miniature coaxial cable and are rated for use up to 18 GHz. Handheld transceivers often use SMA connectors for attaching antennas.

# Practical Circuits

Your General class exam (Element 3) will consist of 35 questions taken from the General class question pool as prepared by the Volunteer Examiner Coordinators' Questions Pool Committee. A certain number of questions are taken from each of the 10 subelements. There will be 3 questions from the subelement shown in this chapter. These questions are divided into 3 groups, labeled G7A through G7C.

## SUBELEMENT G7 — PRACTICAL CIRCUITS
### [3 Exam Questions — 3 Groups]

### G7A — Power supplies; schematic symbols

**G7A01**  **What safety feature does a power-supply bleeder resistor provide?**

A. It acts as a fuse for excess voltage
B. It discharges the filter capacitors
C. It removes shock hazards from the induction coils
D. It eliminates ground-loop current

**(B)**  A bleeder resistor connected across a filter capacitor discharges the capacitor when power is not supplied to the circuit. This minimizes the risk of electrical shock if the supply enclosure is opened, exposing the capacitor terminals.

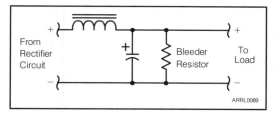

From
Rectifier
Circuit

Bleeder
Resistor

To
Load

ARRL0089

**Figure G7-1 — A bleeder resistor is a safety feature that discharges a power supply filter capacitor when the supply is turned off.**

**G7A02** Which of the following components are used in a power-supply filter network?

A. Diodes
B. Transformers and transducers
C. Quartz crystals
D. Capacitors and inductors

**(D)** A power supply filter network consists of capacitors and inductors used to smooth out the pulses of voltage and current from the rectifier. Capacitors oppose changes in voltage while the inductors oppose changes in current. The combination results in a constant dc output voltage from the supply.

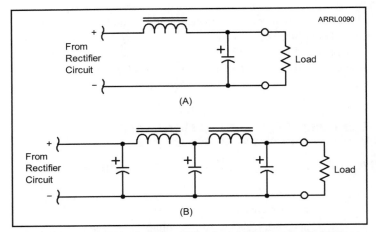

**Figure G7-2 — Part A shows a choke-input power-supply filter circuit. Part B shows a capacitor-input, multisection filter.**

**G7A03** What is the peak-inverse-voltage across the rectifiers in a full-wave bridge power supply?

A. One-quarter the normal output voltage of the power supply
B. Half the normal output voltage of the power supply
C. Double the normal peak output voltage of the power supply
D. Equal to the normal peak output voltage of the power supply

**(D)** In the full-wave center-tapped rectifier circuit, when a rectifier diode is not conducting it must withstand not only the negative peak voltage from its own half of the winding but also the positive voltage from the other winding. Thus the peak inverse voltage applied to the diodes is twice the normal peak output voltage of the supply. In a bridge rectifier circuit the applied PIV to either rectifier is only the peak output voltage of the supply.

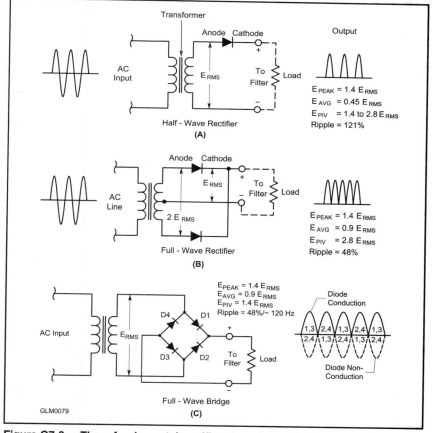

Figure G7-3 — Three fundamental rectifier circuits and the characteristics of their output voltage. (A) Half-wave. (B) Full-wave center-tapped. (C) Full-wave bridge. The half-wave rectifier circuit converts only one-half of the input waveform cycle (180°) while the full-wave circuits convert the entire cycle (360°). In most power supplies, a capacitor is connected across the output of the rectifier and will charge to a voltage of $E_{PEAK}$, the normal peak output of the supply.

**G7A04**  What is the peak inverse voltage across the rectifier in a half-wave power supply?

A. One-half the normal peak output voltage of the power supply
B. One-half the normal output voltage of the power supply
C. Equal to the normal output voltage of the power supply
D. Two times the normal peak output voltage of the power supply

**(D)** The peak inverse voltage applied to the rectifier diode in a half-wave rectifier circuit is twice the supply's peak output voltage because the filter capacitor charges to the peak transformer voltage during the half-cycle when the diode is conducting. During the half-cycle when the diode is not conducting, the capacitor remains charged while the transformer output voltage reaches a peak value of the opposite polarity. Thus the peak inverse voltage is twice the normal peak output voltage of the supply.

**G7A05**  What portion of the AC cycle is converted to DC by a half-wave rectifier?

A. 90 degrees
B. 180 degrees
C. 270 degrees
D. 360 degrees

**(B)** Since there are 360 degrees in a full cycle of ac, a half-wave rectifier converts 180 degrees of the ac input waveform to dc.

**G7A06**  What portion of the AC cycle is converted to DC by a full-wave rectifier?

A. 90 degrees
B. 180 degrees
C. 270 degrees
D. 360 degrees

**(D)** Since there are 360 degrees in a full cycle of ac, a full-wave rectifier converts 360 degrees of the ac input waveform to dc.

**G7A07**  What is the output waveform of an unfiltered full-wave rectifier connected to a resistive load?

A. A series of DC pulses at twice the frequency of the AC input
B. A series of DC pulses at the same frequency as the AC input
C. A sine wave at half the frequency of the AC input
D. A steady DC voltage

**(A)** A full-wave rectifier changes alternating current with positive and negative half cycles into a fluctuating current with all positive pulses. Since the current in this case has not yet been filtered, it is a series of pulses at twice the frequency of the ac input.

**G7A08** Which of the following is an advantage of a switch-mode power supply as compared to a linear power supply?

A. Faster switching time makes higher output voltage possible
B. Fewer circuit components are required
C. High frequency operation allows the use of smaller components
D. All of these choices are correct

**(C)** Switched-mode power supplies operate at a frequency much higher than the 60 Hz ac line current. (Oscillator frequencies of 50 kHz or more are commonly used.) This allows the use of small, lightweight transformers. While the transformer in a linear power supply capable of supplying 20 amperes might weigh 15 or 20 pounds, the transformer for a switched-mode power supply with a similar current rating might weigh 1 or 2 pounds! Switched-mode power supplies have more complex circuits than linear supplies and generally require more components than a simple linear supply.

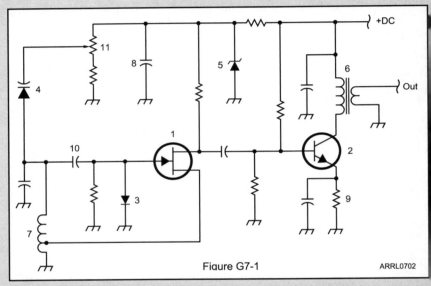

Figure G7-1

ARRL0702

**Question Pool Figure G7-1 —This schematic is used in General class exam for questions G7A09 to G7A13.**

**G7A09**  Which symbol in figure G7-1 represents a field
effect transistor?

A. Symbol 2
B. Symbol 5
C. Symbol 1
D. Symbol 4

**(C)**


**G7A10**  Which symbol in figure G7-1 represents a Zener diode?

A. Symbol 4
B. Symbol 1
C. Symbol 11
D. Symbol 5

**(D)**


**G7A11**  Which symbol in figure G7-1 represents an NPN
junction transistor?

A. Symbol 1
B. Symbol 2
C. Symbol 7
D. Symbol 11

**(B)**


**G7A12**  Which symbol in Figure G7-1 represents a
multiple-winding transformer?

A. Symbol 4
B. Symbol 7
C. Symbol 6
D. Symbol 1

**(C)**


**G7A13**  Which symbol in Figure G7-1 represents a tapped inductor?

A. Symbol 7
B. Symbol 11
C. Symbol 6
D. Symbol 1

**(A)**

## G7B    Digital circuits; amplifiers and oscillators

**G7B01**    Complex digital circuitry can often be replaced by what type of integrated circuit?

A. Microcontroller
B. Charge-coupled device
C. Phase detector
D. Window comparator

**(A)** Microcontrollers are a special type of microprocessor designed to interact with external sensors and circuits to control or operate a piece of equipment, such as an automatic antenna tuner or electronic keyer.

**G7B02**    Which of the following is an advantage of using the binary system when processing digital signals?

A. Binary "ones" and "zeros" are easy to represent with an "on" or "off" state
B. The binary number system is most accurate
C. Binary numbers are more compatible with analog circuitry
D. All of these choices are correct

**(A)** Binary numbers are easily represented by circuits that have two stable states — ON and OFF. The states can be used to represent either a binary 0 or 1 as the designer chooses.

**G7B03** Which of the following describes the function of a two input AND gate?

A. Output is high when either or both inputs are low
B. Output is high only when both inputs are high
C. Output is low when either or both inputs are high
D. Output is low only when both inputs are high

**(B)** If both inputs to an AND gate are true or "1", the AND function output is also "1", otherwise the output is "0".

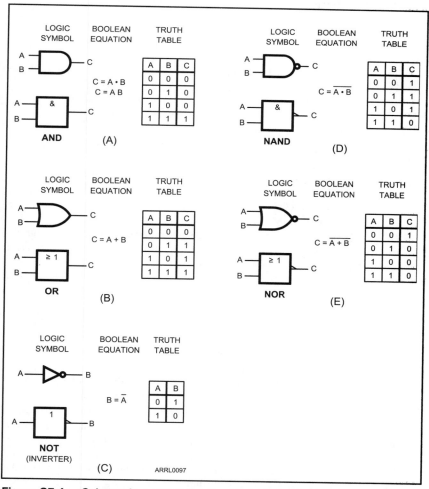

Figure G7-4 — Schematic symbols for the basic digital logic functions with the logic equations and truth tables that describe their operation. The two-input AND gate is shown at A and a two-input NOR gate at E.

**G7B04** **Which of the following describes the function of a two input NOR gate?**

A. Output is high when either or both inputs are low
B. Output is high only when both inputs are high
C. Output is low when either or both inputs are high
D. Output is low only when both inputs are high

**(C)** A NOR (NOT-OR) gate consists of an OR gate with its output inverted. If either or both of the inputs to a two-input NOR gate are true or "1", the OR function is also "1", which is then inverted to "0" at the output.

**G7B05** **How many states does a 3-bit binary counter have?**

A. 3
B. 6
C. 8
D. 16

**(C)** There are 8 states because the number of states in a binary counter is $2^N$, where N is the number of bits in the counter. A 2-bit counter has $2^2 = 4$ states, a 3-bit counter $2^3 = 8$ states, a 4-bit counter $2^4 = 16$ states, and so forth.

**G7B06** **What is a shift register?**

A. A clocked array of circuits that passes data in steps along the array
B. An array of operational amplifiers used for tri state arithmetic operations
C. A digital mixer
D. An analog mixer

**(A)** A shift register consists of a sequence of flip-flop circuits connected output-to-input and sharing a common clock input signal. With each pulse of the clock signal, the state of the input signal (0 or 1) to the shift register is transferred to the output of the first flip-flop and each subsequent flip-flop's output is passed to the next flip-flop's input.

**G7B07** **What are the basic components of virtually all sine wave oscillators?**

A. An amplifier and a divider
B. A frequency multiplier and a mixer
C. A circulator and a filter operating in a feed-forward loop
D. A filter and an amplifier operating in a feedback loop

**(D)** To make an oscillator requires an amplifier and a circuit that routes some of the amplifier's output signal back to the input (called a "feedback loop") such that it is reinforced in the amplifier output. This is called positive feedback. A filter in the feedback loop is used so only signals at the desired frequency are reinforced. Starting with random noise, the oscillator gradually builds up an output signal at the filter frequency until it is self-sustaining.

**G7B08** How is the efficiency of an RF power amplifier determined?

A. Divide the DC input power by the DC output power
B. Divide the RF output power by the DC input power
C. Multiply the RF input power by the reciprocal of the RF output power
D. Add the RF input power to the DC output power

**(B)** Efficiency is defined as the total output power divided by the total input power and is measured in percent. For an RF amplifier, total output power is measured by a wattmeter. The total input power is the dc power required for the amplifier to operate. For example, if to produce 1200 watts of RF output the amplifier requires 1000 mA of plate current at a voltage of 2000 V, efficiency = 1200 watts / (1 A × 2000 V) = 1200 / 2000 = 60%.

**G7B09** What determines the frequency of an LC oscillator?

A. The number of stages in the counter
B. The number of stages in the divider
C. The inductance and capacitance in the tank circuit
D. The time delay of the lag circuit

**(C)** An LC oscillator uses a parallel LC circuit (a tank circuit) as the filter in the feedback loop. (See also the discussion for question G7B07.)

**G7B10** Which of the following is a characteristic of a Class A amplifier?

A. Low standby power
B. High Efficiency
C. No need for bias
D. Low distortion

**(D)** Class A amplifiers are used when low distortion is required. Class A means that current flows in the amplifying device — whether a transistor or vacuum tube — at all times. Because the amplifier is never saturated or cut off, its operation can be optimized for linear reproduction of the input signal. The tradeoff is that Class A amplifiers consume power 100% of the time, require a source of bias to keep them on in the absence of an input signal, and aren't particularly efficient.

**G7B11** For which of the following modes is a Class C power stage appropriate for amplifying a modulated signal?

A. SSB
B. CW
C. AM
D. All of these choices are correct

**(B)** A Class C amplifier conducts current during less than half of the input signal cycle, resulting in high distortion. This rules out Class C amplifiers for any form of amplitude modulation, such as SSB or AM. Class C amplifiers can be used for CW since that mode requires only the presence or absence of a signal. Similarly, Class C is suitable for FM signals that only depend on signal frequency, which is not changed by the amplifier. Class C amplifiers generate significant harmonics, so they require filtering for use as transmitter outputs.

**G7B12** **Which of these classes of amplifiers has the highest efficiency?**

    A. Class A
    B. Class B
    C. Class AB
    D. Class C

**(D)** Class C amplifiers act a lot like switches — they turn on for a fraction of the input signal's cycle and stay off the rest of the time. While these amplifiers are not linear at all and require filters to eliminate the harmonics in their output, they are quite efficient.

**G7B13** **What is the reason for neutralizing the final amplifier stage of a transmitter?**

    A. To limit the modulation index
    B. To eliminate self-oscillations
    C. To cut off the final amplifier during standby periods
    D. To keep the carrier on frequency

**(B)** Neutralization of a power amplifier is a technique that minimizes or cancels the effects of positive feedback. Positive feedback occurs when the output signal is fed back to the input in phase with the input signal, creating an oscillator. Neutralization consists of feeding a portion of the amplifier output back to the input, 180 degrees out of phase with the input, to cancel the positive feedback. This is called negative feedback. Self-oscillation creates powerful spurious signals that cause interference. Self-oscillations can also be sufficiently powerful to damage the amplifier.

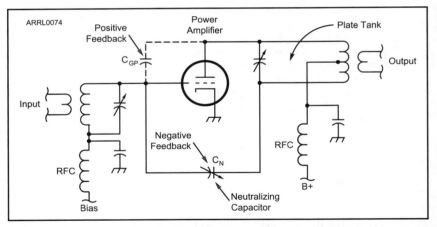

Figure G7-5 — The schematic circuit for an amplifier shows both the source of positive feedback that can create self-oscillation and negative feedback that cancels or neutralizes the positive feedback. Positive feedback is created by the capacitance from the plate to the grid, shows as $C_{GP}$ on the schematic. The amount of capacitance is small, so the frequency of self-oscillation is usually at VHF or in the upper HF range. Capacitor $C_N$ is connected to a point in the output circuit at which the signal has the opposite phase, so the feedback to the grid acts as negative feedback by providing an equal-and-opposite feedback signal to that from $C_{GP}$.

**G7B14** Which of the following describes a linear amplifier?

A. Any RF power amplifier used in conjunction with an amateur transceiver
B. An amplifier in which the output preserves the input waveform
C. A Class C high efficiency amplifier
D. An amplifier used as a frequency multiplier

**(B)** A linear amplifier is defined as one whose output waveform is a copy of the input waveform, although larger in amplitude. Hams refer to power amplifiers as "linears", whether they are operating linearly (for AM or SSB modes) or not (for CW or FM). It is important to understand when linear operation is important. An amplifier designed for FM will not be suitable as an SSB amplifier, for example.

## G7C    Receivers and transmitters; filters, oscillators

**G7C01** Which of the following is used to process signals from the balanced modulator and send them to the mixer in a single-sideband phone transmitter?

A. Carrier oscillator
B. Filter
C. IF amplifier
D. RF amplifier

**(B)** In a single-sideband transmitter, the modulating audio is added to the RF signal in the balanced modulator. The balanced modulator also balances out or cancels the original carrier signal, leaving a double-sideband, suppressed-carrier signal. A filter then removes one of the sidebands, leaving a single-sideband signal that is sent to the mixer, where it combines with the signal from a local oscillator (LO) to produce the RF signal that is amplified and sent to the antenna. The LO shown in the figure is crystal-controlled for fixed-frequency operation. Replacing the crystal-controlled LO with a variable-frequency oscillator (VFO) results in a tunable transmitter similar to those in most modern transceivers.

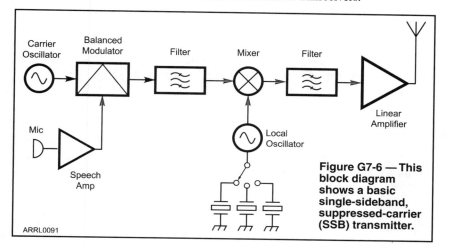

Figure G7-6 — This block diagram shows a basic single-sideband, suppressed-carrier (SSB) transmitter.

ARRL0091

**G7C02**  Which circuit is used to combine signals from the carrier oscillator and speech amplifier and send the result to the filter in a typical single-sideband phone transmitter?

A. Discriminator
B. Detector
C. IF amplifier
D. Balanced modulator

**(D)**  In a single-sideband transmitter, the modulating audio is added to the RF signal in the balanced modulator. The balanced modulator also balances out or cancels the original carrier signal, leaving a double-sideband, suppressed-carrier signal. A filter then removes one of the sidebands, leaving a single-sideband signal that is sent to the mixer, where it combines with the signal from a Local Oscillator (LO) to produce the RF signal that is amplified and sent to the antenna. The LO shown in the figure is crystal-controlled for fixed-frequency operation. Replacing the crystal-controlled LO with a variable-frequency oscillator (VFO) results in a tunable transmitter similar to those in most modern transceivers.

**G7C03**  What circuit is used to process signals from the RF amplifier and local oscillator and send the result to the IF filter in a superheterodyne receiver?

A. Balanced modulator
B. IF amplifier
C. Mixer
D. Detector

**(C)**  In a superheterodyne receiver, the mixer combines signals from the RF amplifier and the local oscillator (LO) then sends those signals to the IF filter, which passes the desired range of frequencies while rejecting the signals at higher and lower frequencies.

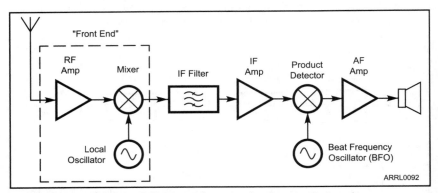

**Figure G7-7 — This block diagram shows a simple superheterodyne SSB receiver.**

**G7C04**  What circuit is used to combine signals from the IF amplifier and BFO and send the result to the AF amplifier in a single-sideband receiver?

A. RF oscillator
B. IF filter
C. Balanced modulator
D. Product detector

**(D)**  In a superheterodyne receiver, the product detector circuit recovers the modulating audio signal by combining the output of the intermediate frequency (IF) amplifier and the beat frequency oscillator (BFO). The recovered audio signal is then passed to the audio frequency (AF) amplifier.

**G7C05**  Which of the following is an advantage of a transceiver controlled by a direct digital synthesizer (DDS)?

A. Wide tuning range and no need for band switching
B. Relatively high power output
C. Relatively low power consumption
D. Variable frequency with the stability of a crystal oscillator

**(D)**  The direct digital synthesizer or DDS replaces the analog VFO circuits with a digital circuit that creates a sine wave as a series of small steps. The duration and amplitude of each step is precisely controlled based on a crystal oscillator. This allows a DDS to act as an oscillator with stability that is comparable to that of a crystal oscillator, while still being adjustable over a wide range.

**G7C06**  What should be the impedance of a low-pass filter as compared to the impedance of the transmission line into which it is inserted?

A. Substantially higher
B. About the same
C. Substantially lower
D. Twice the transmission line impedance

**(B)**  To prevent unwanted reflected power and elevated SWR, keep all elements of the feed line and antenna system at the same impedance. A low-pass filter, designed to be installed at the transmitter output, should have the same impedance as the transmission line.

**G7C07**  What is the simplest combination of stages that implement a superheterodyne receiver?

A. RF amplifier, detector, audio amplifier
B. RF amplifier, mixer, IF discriminator
C. HF oscillator, mixer, detector
D. HF oscillator, pre-scaler, audio amplifier

**(C)**  By definition, a superheterodyne receiver must contain a mixer and a local oscillator. One additional stage, a detector, is necessary to recover the modulating audio. Thus, the simplest superheterodyne consists of a mixer, oscillator, and detector. Practical receivers add more amplifiers to improve sensitivity and filters to reject unwanted signals.

**G7C08** What type of circuit is used in many FM receivers to convert signals coming from the IF amplifier to audio?

A. Product detector
B. Phase inverter
C. Mixer
D. Discriminator

**(D)** A frequency discriminator converts the variations in frequency of the limiter's output signal into amplitude variations, recovering the modulating audio signal. A quadrature detector is another type of FM demodulation circuit that converts changes in frequency into amplitude variations.

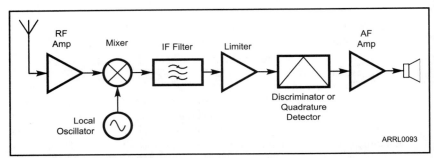

**Figure G7-8 — This block diagram shows a simple FM receiver.**

**G7C09** Which of the following is needed for a Digital Signal Processor IF filter?

A. An analog to digital converter
B. A digital to analog converter
C. A digital processor chip
D. All of the these choices are correct

**(D)** Digital Signal Processing (DSP) is the process of converting analog signals to digital data, processing them with software programs, then converting the signals back to analog form. Digital noise reduction is but one application — filtering, demodulation, and decoding are also commonly performed by DSP circuits. (See also the discussion for questions G4C11 and G4C12.)

**G7C10** How is Digital Signal Processor filtering accomplished?

A. By using direct signal phasing
B. By converting the signal from analog to digital and using digital processing
C. By differential spurious phasing
D. By converting the signal from digital to analog and taking the difference of mixing products

**(B)** See G7C09.

**G7C11** What is meant by the term "software defined radio" (SDR)?

A. A radio in which most major signal processing functions are performed by software
B. A radio which provides computer interface for automatic logging of band and frequency
C. A radio which uses crystal filters designed using software
D. A computer model which can simulate performance of a radio to aid in the design process

**(A)** In a software-defined radio (SDR) nearly all of the radio's functions are performed by digital hardware. This allows the radio's operation to be changed and controlled by software without having to change the way in which the radio is physically constructed.

# Signals and Emissions

Your General class exam (Element 3) will consist of 35 questions taken from the General class question pool as prepared by the Volunteer Examiner Coordinators' Questions Pool Committee. A certain number of questions are taken from each of the 10 subelements. There will be 2 questions from the subelement shown in this chapter. These questions are divided into 2 groups, labeled G8A and G8B.

## SUBELEMENT G8 — SIGNALS AND EMISSIONS
## [2 Exam Questions — 2 Groups]

### G8A Carriers and modulation: AM; FM; single and double sideband; modulation envelope; overmodulation

**G8A01** What is the name of the process that changes the envelope of an RF wave to carry information?

A. Phase modulation
B. Frequency modulation
C. Spread spectrum modulation
D. Amplitude modulation

**(D)** There are several ways to change or modulate an RF wave for the purpose of conveying information. CW simply turns the signal on and off to convey information. Amplitude modulation changes the strength or amplitude of the wave while frequency modulation changes its frequency. Phase modulation conveys information by changing the phase of the wave.

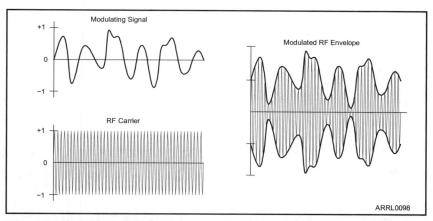

**Figure G8-1 — This drawing shows the relationship between the modulating audio waveform, the RF carrier and the resulting RF envelope in a double-sideband, full-carrier amplitude-modulated signal.**

**G8A02** What is the name of the process that changes the phase angle of an RF wave to convey information?

A. Phase convolution
B. Phase modulation
C. Angle convolution
D. Radian inversion

**(B)** See question G8A01.

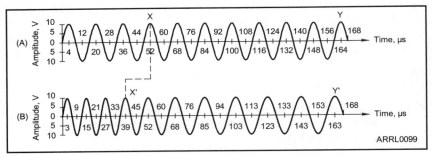

Figure G8-2 — This drawing shows a graphical representation of phase modulation. The unmodulated wave is shown at A. Part B shows the modulated wave. After modulation, cycle X' occurs earlier than cycle X did. All the cycles to the left of X' are compressed, and to the right they are spread out.

**G8A03** What is the name of the process which changes the frequency of an RF wave to convey information?

A. Frequency convolution
B. Frequency transformation
C. Frequency conversion
D. Frequency modulation

**(D)** See question G8A01.

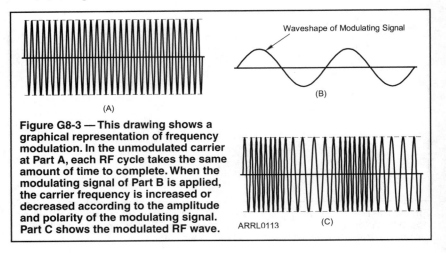

Figure G8-3 — This drawing shows a graphical representation of frequency modulation. In the unmodulated carrier at Part A, each RF cycle takes the same amount of time to complete. When the modulating signal of Part B is applied, the carrier frequency is increased or decreased according to the amplitude and polarity of the modulating signal. Part C shows the modulated RF wave.

**G8A04** What emission is produced by a reactance modulator connected to an RF power amplifier?

A. Multiplex modulation
B. Phase modulation
C. Amplitude modulation
D. Pulse modulation

**(B)** Reactance modulators are the most common method of generating phase modulation (PM) signals. Phase modulators cause the output signal phase to vary with both the modulating signal's amplitude and frequency. (See also the discussion for question G8A01.)

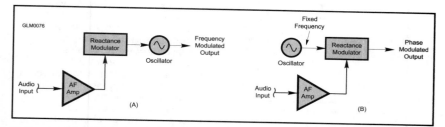

**Figure G8-4 — Reactance modulators can be used to create frequency modulation (A) or phase modulation (B).**

**G8A05** What type of modulation varies the instantaneous power level of the RF signal?

A. Frequency shift keying
B. Pulse position modulation
C. Frequency modulation
D. Amplitude modulation

**(D)** In an AM transmission, at any given instant the amplitude or envelope of the RF signal changes according to the modulating audio signal. (See also the discussion for question G8A01.)

**G8A06**  What is one advantage of carrier suppression in a single-sideband phone transmission?

A. Audio fidelity is improved
B. Greater modulation percentage is obtainable with lower distortion
C. The available transmitter power can be used more effectively
D. Simpler receiving equipment can be used

**(C)** In a single-sideband, suppressed-carrier, amplitude-modulated transmitter, the RF carrier signal is reduced or suppressed. The less energy that is used by the carrier the more power that can be put into the sidebands. This increases the efficiency of the system.

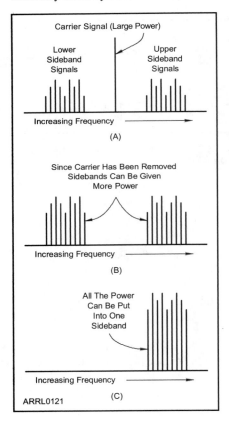

**Figure G8-5 — This drawing shows the frequency spectrum of an amplitude-modulated radio signal. Part A shows a double-sideband, full-carrier signal. When the carrier is removed, only the two sidebands are left, as shown in Part B. When one sideband is suppressed, as shown in Part C, the full transmitter power can be concentrated in one sideband.**

**G8A07** Which of the following phone emissions uses the narrowest frequency bandwidth?

A. Single sideband
B. Double sideband
C. Phase modulation
D. Frequency modulation

**(A)** In a single-sideband (SSB) amplitude-modulated transmitter, the RF carrier signal is reduced or suppressed, and one of the sidebands is removed. Suppression of the carrier allows more power to be put into the sideband. Because the carrier is suppressed and one sideband is filtered out, only enough frequency bandwidth is required to transmit a single sideband. This has the narrowest bandwidth of all the popular phone emissions. The bandwidth of an SSB signal is between about 2 and 3 kHz, the bandwidth of a double-sideband AM signal is about 6 kHz and the bandwidth of frequency and phase modulated phone signals is about 16 kHz.

**G8A08** **Which of the following is an effect of over-modulation?**

   A. Insufficient audio
   B. Insufficient bandwidth
   C. Frequency drift
   D. Excessive bandwidth

**(D)** When an SSB signal is overmodulated the output waveform of the signal is distorted, which causes spurious emissions outside the normal bandwidth of the signal. When an FM or PM signal is overmodulated, the deviation of the signal becomes too high and again, spurious emissions appear outside the normal bandwidth of the signal. In both cases, the spurious emissions cause interference to other stations.

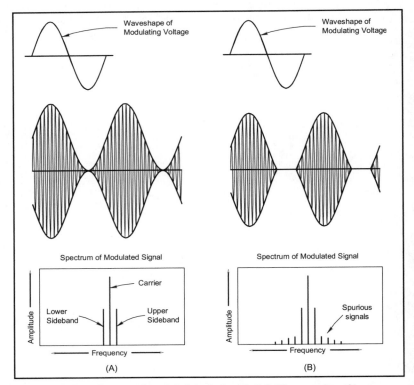

**Figure G8-6 — A properly modulated signal at A. The results of over-modulation are visible at B. This distorted signal causes interference on nearby frequencies.**

**G8A09** What control is typically adjusted for proper ALC setting on an amateur single sideband transceiver?

A. The RF clipping level
B. Transmit audio or microphone gain
C. Antenna inductance or capacitance
D. Attenuator level

**(B)** To reduce overmodulation, the Automatic Level Control (ALC) circuit reduces microphone gain when it detects excessive audio levels. For proper adjustment on most transmitters, the microphone gain control should be adjusted so that there is a slight movement of the ALC meter on modulation peaks.

**G8A10** What is meant by flat-topping of a single-sideband phone transmission?

A. Signal distortion caused by insufficient collector current
B. The transmitter's automatic level control is properly adjusted
C. Signal distortion caused by excessive drive
D. The transmitter's carrier is properly suppressed

**(C)** The figure shows an overmodulated signal as seen on an oscilloscope with flattening at the maximum levels of the envelope. This is referred to as flat-topping.

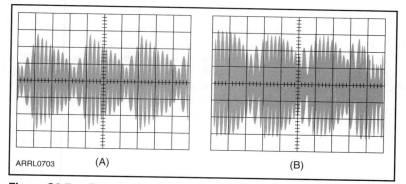

ARRL0703     (A)                              (B)

Figure G8-7 — Part A shows the waveform of a properly adjusted SSB transmitter. Part B shows a severely clipped, distorted signal.

**G8A11** What happens to the RF carrier signal when a modulating audio signal is applied to an FM transmitter?

A. The carrier frequency changes proportionally to the instantaneous amplitude of the modulating signal
B. The carrier frequency changes proportionally to the amplitude and frequency of the modulating signal
C. The carrier amplitude changes proportionally to the instantaneous frequency of the modulating signal
D. The carrier phase changes proportionally to the instantaneous amplitude of the modulating signal

**(A)** Frequency Modulation (FM) results when the modulating signal causes the frequency of an oscillator to change in proportion to the modulating signal's amplitude. If the oscillator's frequency changes in proportion to both the modulating signal's amplitude and frequency, the result is phase modulation (PM).

**G8A12** What signal(s) would be found at the output of a properly adjusted balanced modulator?

A. Both upper and lower sidebands
B. Either upper or lower sideband, but not both
C. Both upper and lower sidebands and the carrier
D. The modulating signal and the unmodulated carrier

**(A)** The process of mixing two ac signals creates two new signals. One of the new signals has a frequency equal to the sum of the two input signal frequencies and the other new signal has a frequency equal to the difference between the two input signal frequencies. The output signal from most mixers will also include some signal energy at the original input signals. A mixer can generate a double-sideband signal if the input signals are the carrier and the modulating audio.

A balanced modulator is a mixer that includes some additional circuitry that can be adjusted to balance out or cancel the unwanted signals that remain at the carrier and modulating audio frequencies. When a balanced modulator is properly adjusted, neither of the original input signals reach the modulator output, only the sum and difference frequency signals. These signals are the upper (sum) and lower (difference) sidebands of an AM signal. A balanced modulator is used to produce the modulated RF signal for a single-sideband transmitter. The unwanted sideband is removed by a filter to produce the SSB signal.

## G8B Frequency mixing; multiplication; HF data communications; bandwidths of various modes; deviation

**G8B01** What receiver stage combines a 14.250 MHz input signal with a 13.795 MHz oscillator signal to produce a 455 kHz intermediate frequency (IF) signal?

A. Mixer
B. BFO
C. VFO
D. Discriminator

**(A)** The mixer stage in a receiver combines the input signal with an oscillator signal to produce the sum and the difference of the two signals. Filters or mixer design will give one desired new signal, known as the intermediate frequency (IF) signal. For example, if a 13.795 MHz variable frequency oscillator (VFO) signal is mixed with a 14.25 MHz RF signal, it will produce new signals at 28.045 MHz and 0.455 MHz, or 455 kHz. A filter will remove the 28.045 MHz signal and pass the 455 kHz IF signal.

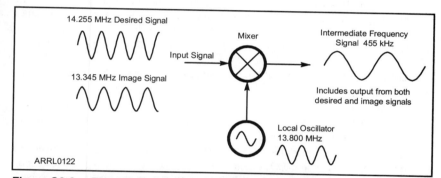

**Figure G8-8 — The mixer stage in a receiver combines an RF input signal with a local oscillator (LO) signal to produce an intermediate frequency (IF) signal.**

**G8B02** If a receiver mixes a 13.800 MHz VFO with a 14.255 MHz received signal to produce a 455 kHz intermediate frequency (IF) signal, what type of interference will a 13.345 MHz signal produce in the receiver?

A. Quadrature noise
B. Image response
C. Mixer interference
D. Intermediate interference

**(B)** The mixer stage in a receiver combines the input signal with an oscillator signal to produce the sum and the difference of the two signals. Filters or mixer design will select the desired new signal, known as the intermediate frequency (IF) signal. In this case the input signal is 14.255 MHz and the oscillator signal is 13.800 MHz. The sum of these two signals is 28.055 MHz and the difference, which is used as the intermediate frequency, is 0.455 MHz or 455 kHz. These signals are then passed through a filter to eliminate the unwanted frequencies leaving only the desired intermediate frequency of 455 kHz.

A signal at 13.455 MHz, when subtracted from the 13.800 MHz oscillator signal will also produce the 455 kHz intermediate frequency. (See Figure G8-8.) When an undesired input signal also produces a signal at the intermediate frequency, the resulting interference is called image response interference. One way to reduce image response interference is to use an input filter that allows signals in the desired receive frequency range to pass through, but blocks signals outside of that range.

**G8B03** What is another term for the mixing of two RF signals?

A. Heterodyning
B. Synthesizing
C. Cancellation
D. Phase inverting

**(A)** The process of mixing signals is known as heterodyning. A receiver that mixes a local oscillator (LO) signal with a received RF signal to produce a signal at some intermediate frequency (IF) that is higher in frequency than the baseband audio signal is called a superheterodyne receiver, or a superhet. The IF signal is processed further by filtering and amplifying it, and then converting the signal to baseband audio. The audio signal is further amplified and fed to a speaker or is connected to headphones so you can hear the received signal.

**G8B04** What is the name of the stage in a VHF FM transmitter that generates a harmonic of a lower frequency signal to reach the desired operating frequency?

A. Mixer
B. Reactance modulator
C. Pre-emphasis network
D. Multiplier

**(D)** An FM transmitter often makes use of a device called a frequency multiplier. In a VHF FM transmitter, a reactance modulator operates on a radio-frequency oscillator that is normally operating in the high-frequency (HF) range. A frequency multiplier doubles or triples the frequency of the modulated signal, usually by generating harmonics of the modulated HF signal and selecting one for output to the next stage in the transmitter. Sometimes several multiplier stages are required to produce a signal at the desired output frequency."

**G8B05** Why isn't frequency modulated (FM) phone used below 29.5 MHz?

A. The transmitter efficiency for this mode is low
B. Harmonics could not be attenuated to practical levels
C. The wide bandwidth is prohibited by FCC rules
D. The frequency stability would not be adequate

**(C)** FM phone signals have a bandwidth of about 16 kHz. At frequencies below 29.5 MHz, the maximum FCC allowable bandwidths are narrower than 16 kHz. The HF bands are relatively narrow, and would not be able to accommodate many wide bandwidth FM signals.

**G8B06** **What is the total bandwidth of an FM-phone transmission having a 5 kHz deviation and a 3 kHz modulating frequency?**

    A. 3 kHz
    B. 5 kHz
    C. 8 kHz
    D. 16 kHz

**(D)** To determine bandwidth of an FM-phone transmission, use the following formula:

$$BW = 2 \times (D + M)$$

where:

    BW = bandwidth
    D = frequency deviation (the instantaneous change in frequency for a given signal)
    M = maximum modulating audio frequency

The total bandwidth of an FM-phone transmission having a 5 kHz deviation and a 3 kHz modulating frequency would be:

$$2 \times (5\ kHz + 3\ kHz) = 16\ kHz$$

**G8B07** **What is the frequency deviation for a 12.21-MHz reactance-modulated oscillator in a 5-kHz deviation, 146.52-MHz FM-phone transmitter?**

    A. 101.75 Hz
    B. 416.7 Hz
    C. 5 kHz
    D. 60 kHz

**(B)** An FM transmitter often makes use of a device called a frequency multiplier. In a VHF FM transmitter, a reactance modulator operates on a radio-frequency oscillator that is normally operating in the high-frequency (HF) range. A frequency multiplier doubles or triples the frequency of the modulated signal, usually by generating harmonics of the modulated HF signal and selecting one for output to the next stage in the transmitter. Sometimes several multiplier stages are required to produce a signal at the desired output frequency. When an FM signal's frequency is multiplied, both the carrier frequency and the amount of deviation are both multiplied. This is why the deviation of the modulated oscillator is less than that of the final output signal.

To determine the oscillator frequency deviation, divide the output frequency by the oscillator frequency to determine the multiplication factor of the transmitter. Then divide the desired output deviation by the multiplication factor. The frequency deviation for a 12.21 MHz reactance-modulated oscillator in a 5 kHz deviation, 146.52 MHz FM-phone transmitter is 416.7 Hz.

$$\text{Multiplication Factor} = \frac{\text{Transmitter Frequency}}{\text{HF Oscillator Frequency}} = \frac{146.52\ MHz}{12.21\ MHz} = 12$$

Then:

$$\text{Desired Oscillator Deviation} = \frac{\text{Transmitter Deviation}}{\text{Multiplication Factor}} = \frac{5000\ Hz}{12} = 416.7\ Hz$$

**G8B08**  Why is it important to know the duty cycle of the data mode you are using when transmitting?

A. To aid in tuning your transmitter
B. Some modes have high duty cycles which could exceed the transmitter's average power rating.
C. To allow time for the other station to break in during a transmission
D. All of these choices are correct

**(B)** Most Amateur Radio transmitters are not designed to operate at full power output for an extended time. The final output stage is not able to dissipate all of the excess heat that is generated when the transmitter is producing full output power continuously. When you are operating CW, for example, the transmitter is turned on and off to form the Morse code characters so that the transmitter is only operating at full power about 40 to 50% of the time. During the off times, the amplifier stage cools sufficiently to allow full-power operation. When you are operating single-sideband voice, the transmitter is producing full power only when your voice reaches maximum amplitude. For a typical SSB conversation, the transmitter is operating at full power only about 20 to 25% of the time.

When you are operating some data modes, however, your transmitter may be operating at full power the entire time you are transmitting. For Baudot radioteletype the transmitter is producing full output power, switching between the mark and space tones of the code, so the duty cycle is 100%. For PSK31 and similar modes, the transmitter is producing full power for virtually the entire transmit time, so the duty cycle is 100%. PACTOR, packet radio and a few other modes have slightly reduced duty cycles because the transmitter sends some data and then waits to receive an acknowledgement. If you are operating a high-duty-cycle mode you should reduce your transmit power to prevent overheating the amplifier. Reduce your transmitter power to about 50% of maximum output power for most data modes.

**G8B09**  Why is it good to match receiver bandwidth to the bandwidth of the operating mode?

A. It is required by FCC rules
B. It minimizes power consumption in the receiver
C. It improves impedance matching of the antenna
D. It results in the best signal to noise ratio

**(D)** By matching the receiver bandwidth and the signal bandwidth, noise outside the signal's bandwidth is rejected and no necessary signal energy is discarded. Both result in improvement of the signal-to-noise ratio (SNR).

**G8B10**  What does the number 31 represent in PSK31?

A. The approximate transmitted symbol rate
B. The version of the PSK protocol
C. The year in which PSK31 was invented
D. The number of characters that can be represented by PSK31

**(A)** PSK31 is a digital mode that transmits symbols at a rate of 31.25 baud.

**G8B11** How does forward error correction allow the receiver to correct errors in received data packets?

A. By controlling transmitter output power for optimum signal strength
B. By using the varicode character set
C. By transmitting redundant information with the data
D. By using a parity bit with each character

**(C)** Forward error correction (FEC) is the practice of sending redundant data in the transmitted packet that allows the receiver to correct some types of errors that may be caused by noise, fading or interference. There are a number of FEC methods involving special codes.

**G8B12** What is the relationship between transmitted symbol rate and bandwidth?

A. Symbol rate and bandwidth are not related
B. Higher symbol rates require higher bandwidth
C. Lower symbol rates require higher bandwidth
D. Bandwidth is constant for data mode signals

**(B)** Advanced modulation techniques can pack multiple bits of data into each transmitted symbol, but it is the symbol rate that sets a minimum limit on bandwidth. Increasing the rate at which symbols are transmitted requires more signal bandwidth in order to maintain a minimum signal-to-noise ratio.

# Antennas and Feed Lines

Your General class exam (Element 3) will consist of 35 questions taken from the General class question pool as prepared by the Volunteer Examiner Coordinators' Questions Pool Committee. A certain number of questions are taken from each of the 10 subelements. There will be 4 questions from the subelement shown in this chapter. These questions are divided into 4 groups, labeled G9A through G9D.

## SUBELEMENT G9 — ANTENNAS AND FEED LINES
[4 Exam Questions — 4 Groups]

### G9A  Antenna feed lines: characteristic impedance and attenuation; SWR calculation, measurement and effects; matching networks

**G9A01**  Which of the following factors determine the characteristic impedance of a parallel conductor antenna feed line?

A. The distance between the centers of the conductors and the radius of the conductors

B. The distance between the centers of the conductors and the length of the line

C. The radius of the conductors and the frequency of the signal

D. The frequency of the signal and the length of the line

**(A)** The characteristic impedance of a parallel-conductor feed line depends on the distance between the conductor centers and the radius of the conductors.

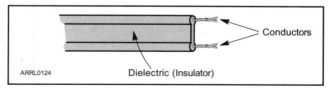

**Figure G9-1 — This drawing shows the construction of common 300-ohm twin lead. This is one form of parallel-conductor feed line**

**G9A02**  What are the typical characteristic impedances of coaxial
cables used for antenna feed lines at amateur stations?

A. 25 and 30 ohms
B. 50 and 75 ohms
C. 80 and 100 ohms
D. 500 and 750 ohms

**(B)** Common coaxial cables used as antenna feed lines have characteristic
impedances of 50 or 75 ohms.

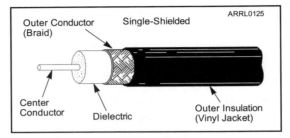

Figure G9-2 — Coaxial cables consist of a
conductor surrounded by insulation. The
second conductor, called the shield, goes
around the insulation. Plastic insulation goes
around the entire cable.

**G9A03**  What is the characteristic impedance of flat ribbon TV
type twinlead?

A. 50 ohms
B. 75 ohms
C. 100 ohms
D. 300 ohms

**(D)** The flat ribbon type of feed line often used with TV antennas has a
characteristic impedance of 300 ohms. This feed line is called twin lead.

**G9A04** What is the reason for the occurrence of reflected power at the point where a feed line connects to an antenna?

A. Operating an antenna at its resonant frequency
B. Using more transmitter power than the antenna can handle
C. A difference between feed-line impedance and antenna feed-point impedance
D. Feeding the antenna with unbalanced feed line

**(C)** Whenever power traveling along a feed line encounters a different impedance from the characteristic impedance of the feed line, such as at an antenna, some of the power is reflected back toward the power source. The greater the difference between the feed line's characteristic impedance and the new impedance, the larger the fraction of power that is reflected.

Power reflected back from an antenna returns to the transmitter, which in turn reflects the power back toward the antenna, creating a standing wave. When the transmitter and antenna impedances are not matched, less power is transferred to the antenna because of the extra loss incurred as the reflected power travels up and down the feed line.

**G9A05** How does the attenuation of coaxial cable change as the frequency of the signal it is carrying increases?

A. It is independent of frequency
B. It increases
C. It decreases
D. It reaches a maximum at approximately 18 MHz

**(B)** Feed line loss is greater at higher frequencies. For example, if you were to use the same type of coaxial cable for your 160 meter antenna as for your 2 meter antenna, there would be much more loss at the higher 2 meter frequencies.

**G9A06** In what values are RF feed line losses usually expressed?

A. ohms per 1000 ft
B. dB per 1000 ft
C. ohms per 100 ft
D. dB per 100 ft

**(D)** RF feed line loss is normally specified in decibels of loss for each 100 feet of line. Loss must also be specified at a certain frequency.

**G9A07** What must be done to prevent standing waves on an antenna feed line?

A. The antenna feed point must be at DC ground potential
B. The feed line must be cut to an odd number of electrical quarter wavelengths long
C. The feed line must be cut to an even number of physical half wavelengths long
D. The antenna feed-point impedance must be matched to the characteristic impedance of the feed line

**(D)** To eliminate reflected power, the antenna impedance must be matched to the characteristic impedance of the feed line. If the impedances are matched, all of the feed line power is transferred to the antenna. The problem with reflected power causing standing waves and raising the feed line SWR is not generally transmitter efficiency but rather the transmitter's reaction to the high SWR. Modern solid-state transmitters usually have protection circuitry that reduces the power output in the presence of high SWR. (When the SWR is high there are high voltages present that can damage components.) By reducing SWR, the transmitter can operate at maximum power output.

**G9A08** If the SWR on an antenna feed line is 5 to 1, and a matching network at the transmitter end of the feed line is adjusted to 1 to 1 SWR, what is the resulting SWR on the feed line?

A. 1 to 1
B. 5 to 1
C. Between 1 to 1 and 5 to 1 depending on the characteristic impedance of the line
D. Between 1 to 1 and 5 to 1 depending on the reflected power at the transmitter

**(B)** A matching network at the transmitter does not change the SWR on the feed line, so the feed line SWR is still 5:1.

**G9A09** What standing wave ratio will result from the connection of a 50-ohm feed line to a non-reactive load having a 200-ohm impedance?

A. 4:1
B. 1:4
C. 2:1
D. 1:2

**(A)** If a load connected to a feed line is purely resistive, the SWR can be calculated by dividing the line characteristic impedance by the load resistance or vice versa, whichever gives a value greater than one. 200 / 50 = 4:1 SWR.

**G9A10**   What standing wave ratio will result from the connection
of a 50-ohm feed line to a non-reactive load having a
10-ohm impedance?

A. 2:1
B. 50:1
C. 1:5
D. 5:1

**(D)**  If a load connected to a feed line is purely resistive, the SWR can be calculated by dividing the line characteristic impedance by the load resistance or vice versa, whichever gives a value greater than one. 50 / 10 = 5:1 SWR.

**G9A11**   What standing wave ratio will result from the connection
of a 50-ohm feed line to a non-reactive load having a 50-ohm
impedance?

A. 2:1
B. 1:1
C. 50:50
D. 0:0

**(B)**  If a load connected to a feed line is purely resistive, the SWR can be calculated by dividing the line characteristic impedance by the load resistance or vice versa, whichever gives a value greater than one. 50 / 50 = 1:1 SWR.

**G9A12**   What would be the SWR if you feed a vertical antenna that
has a 25-ohm feed-point impedance with 50-ohm
coaxial cable?

A. 2:1
B. 2.5:1
C. 1.25:1
D. You cannot determine SWR from impedance values

**(A)**  If a load connected to a feed line is purely resistive, the SWR can be calculated by dividing the line characteristic impedance by the load resistance or vice versa, whichever gives a value greater than one. 50 / 25 = 2:1 SWR.

**G9A13**   What would be the SWR if you feed an antenna that has a
300-ohm feed-point impedance with 50-ohm coaxial cable?

A. 1.5:1
B. 3:1
C. 6:1
D. You cannot determine SWR from impedance values

**(C)**  If a load connected to a feed line is purely resistive, the SWR can be calculated by dividing the line characteristic impedance by the load resistance or vice versa, whichever gives a value greater than one. 300 / 50 = 6:1 SWR.

## G9B    Basic antennas

**G9B01**    What is one disadvantage of a directly fed random-wire antenna?

   A. It must be longer than 1 wavelength
   B. You may experience RF burns when touching metal objects in your station
   C. It produces only vertically polarized radiation
   D. It is not effective on the higher HF bands

**(B)** A random-wire antenna consist of a wire connected directly to the transmitter at one end. It can be of any length because an antenna tuner is used to match the impedances. It does not require a feed line because the single piece of wire serves as both a feed line and an antenna. It is considered to be a multiband antenna because the antenna tuner will match the impedances for several frequency bands. One significant disadvantage of a random-wire antenna is that you may experience RF "hot spots" in your station because the station equipment and ground system are part of your antenna system!

**G9B02**    What is an advantage of downward sloping radials on a quarter wave ground-plane antenna?

   A. They lower the radiation angle
   B. They bring the feed-point impedance closer to 300 ohms
   C. They increase the radiation angle
   D. They bring the feed-point impedance closer to 50 ohms

**(D)** A ground-plane antenna is often constructed with a ¼-wavelength vertical radiating element and four ¼-wavelength horizontal "radial" wires that form the ground plane. You can change the impedance of a ground-plane antenna by changing the angle of the radials. Bending or sloping the radials downward to about a 45-degree angle will increase the impedance from approximately 35 ohms to approximately 50 ohms, reducing SWR.

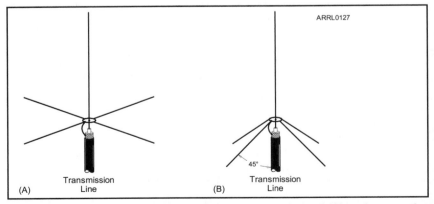

**Figure G9-3 — A ground-plane antenna may use horizontal (A) or downward-sloping radials. By sloping the radials down (B), the feed point impedance is raised closer to 50 ohms to present a better impedance match to 50-ohm coaxial cable.**

**G9B03**  What happens to the feed-point impedance of a ground-plane antenna when its radials are changed from horizontal to downward-sloping?

A. It decreases
B. It increases
C. It stays the same
D. It reaches a maximum at an angle of 45 degrees

**(B)**  See the discussion for question G9B02.

**G9B04**  What is the low angle azimuthal radiation pattern of an ideal half-wavelength dipole antenna installed ½ wavelength high and parallel to the Earth?

A. It is a figure-eight at right angles to the antenna
B. It is a figure-eight off both ends of the antenna
C. It is a circle (equal radiation in all directions)
D. It has a pair of lobes on one side of the antenna and a single lobe on the other side

**(A)**  A ½-wavelength dipole antenna ½ wavelength or more above the ground radiates its signals in a bi-directional fashion with maximum radiation at a 90° angle to the antenna. This is called a "figure 8" radiation pattern. If the antenna is placed less than ½ wavelength above the ground, reflections from the ground will cause more of the antenna's signal to be radiated at high vertical angles and the pattern becomes omnidirectional.

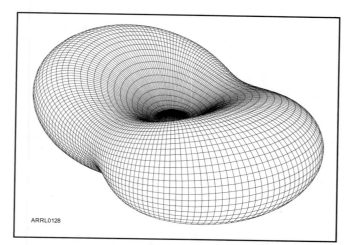

ARRL0128

**Figure G9-4** — This drawing shows the radiation pattern of a ½-wavelength dipole antenna installed ½ wavelength above ground.

**G9B05** How does antenna height affect the horizontal (azimuthal) radiation pattern of a horizontal dipole HF antenna?

A. If the antenna is too high, the pattern becomes unpredictable
B. Antenna height has no effect on the pattern
C. If the antenna is less than ½ wavelength high, the azimuthal pattern is almost omnidirectional
D. If the antenna is less than ½ wavelength high, radiation off the ends of the wire is eliminated

**(C)** As height is reduced below ½ wavelength the antenna pattern becomes almost omnidirectional, sending signals nearly equally in all compass directions.

**G9B06** Where should the radial wires of a ground-mounted vertical antenna system be placed?

A. As high as possible above the ground
B. Parallel to the antenna element
C. On the surface or buried a few inches below the ground
D. At the top of the antenna

**(C)** In most installations, ground conductivity is inadequate to serve as the antenna's ground plane, so an artificial ground screen must be made from wires placed along the ground near the base of the antenna. These radials are usually ¼ wavelength or longer. Depending on ground conductivity, 8, 16, 32 or more radials may be required to form an effective ground. The radial wires of a ground-mounted vertical antenna should be placed on the ground surface or buried a few inches below the surface.

**G9B07** How does the feed-point impedance of a ½ wave dipole antenna change as the antenna is lowered from ¼ wave above ground?

A. It steadily increases
B. It steadily decreases
C. It peaks at about ⅛ wavelength above ground
D. It is unaffected by the height above ground

**(B)** As the antenna is lowered below ¼ wavelength above ground, the impedance steadily decreases to a very low value when placed directly on the ground.

**G9B08** How does the feed-point impedance of a ½ wave dipole change as the feed-point location is moved from the center toward the ends?

A. It steadily increases
B. It steadily decreases
C. It peaks at about ⅛ wavelength from the end
D. It is unaffected by the location of the feed point

**(A)** The center of a ½-wavelength dipole is the location of the lowest feed point impedance, approximately 72 ohms in free space. At the ends of the dipole, feed point impedance is several thousand ohms. In between, feed point impedance increases steadily as the feed point is moved from the center toward the ends of the antenna.

**G9B09** Which of the following is an advantage of a horizontally polarized as compared to vertically polarized HF antenna?

A. Lower ground reflection losses
B. Lower feed-point impedance
C. Shorter Radials
D. Lower radiation resistance

**(A)** The signals from a horizontally polarized antenna have lower losses when reflecting from the ground. This is because the horizontal polarization of the wave induces currents that flow along the surface of the ground. Vertical polarization tends to induce currents that flow vertically in the ground, where losses are higher.

Radio waves reflecting from the ground have lower losses when the polarization of the wave is parallel to the ground. That is, when the waves are horizontally polarized. Because the reflected waves combine with the direct waves (not reflected) to make up the antenna's radiation pattern, lower reflection loss results in stronger signal strength.

Ground-mounted vertical antennas, however, are able to generate stronger signals at low angles of radiation than horizontally polarized antennas at low heights. This means they are often preferred for DX contacts on the lower HF bands where it is impractical to raise horizontally polarized antennas to the height necessary for strong low-angle signals.

**G9B10** What is the approximate length for a ½-wave dipole antenna cut for 14.250 MHz?

A. 8 feet
B. 16 feet
C. 24 feet
D. 32 feet

**(D)** In free space, ½ wavelength in feet equals 492 divided by frequency in MHz. If you cut a piece of wire that length, however, you'll find it is too long to resonate at the desired frequency. A resonant ½-wave dipole made of ordinary wire will be shorter than the free-space wavelength for several reasons. First, the physical thickness of the wire makes it look a bit longer electrically than it is physically. The lower the length-to-diameter (l/d) ratio of the wire, the shorter it will be when it is resonant. Second, the dipole's height above ground also affects its resonant frequency. In addition, nearby conductors, insulation on the wire, the means by which the wire is secured to the insulators and to the feed line also affect the resonant length. For these reasons, a single universal formula for dipole length, such as the common 468/f, is not very useful. You should be start with a length near the free-space length and be prepared to trim the dipole to resonance using an SWR meter or antenna analyzer. The exam only requires that you identify an approximate resonant length for a dipole. Use the free-space length, calculated as 492 / f (in MHz), and select the closest choice. In this case, length (feet) = 492 / 14.250 = 34.5 feet, so select the closest value — 32 feet.

**G9B11** What is the approximate length for a ½-wave dipole antenna cut for 3.550 MHz?

A. 42 feet
B. 84 feet
C. 131 feet
D. 263 feet

**(C)** (See the discussion for G9B10.) Calculate length (feet) = 492 / 3.550 = 139 ft. The closest value is 131 feet.

**G9B12** What is the approximate length for a ¼-wave vertical antenna cut for 28.5 MHz?

A. 8 feet
B. 11 feet
C. 16 feet
D. 21 feet

**(A)** (See the discussion for G9B10.) A ¼-wavelength antenna would be half as long as a ½-wavelength antenna, so calculate length (feet) = 246 / 28.5 = 8.6 ft. The closest value is 8 feet.

## G9C    Directional antennas

**G9C01** Which of the following would increase the bandwidth of a Yagi antenna?

A. Larger diameter elements
B. Closer element spacing
C. Loading coils in series with the element
D. Tapered-diameter elements

**(A)** Using larger diameter elements can increase the SWR bandwidth of a parasitic beam antenna, such as a Yagi antenna. The exact length of the elements becomes less critical when larger diameter elements are used.

**G9C02** What is the approximate length of the driven element of a Yagi antenna?

A. ¼ wavelength
B. ½ wavelength
C. ¾ wavelength
D. 1 wavelength

**(B)** A Yagi antenna consists of a driven element that is close to ½ wavelength long with one or more parasitic elements that help direct the radiated energy in one direction. Directors are parasitic elements that are mounted along the antenna's supporting boom in the preferred direction of radiation and reflectors are mounted in the opposite direction.

**G9C03** Which statement about a three-element, single-band Yagi antenna is true?

A. The reflector is normally the shortest parasitic element
B. The director is normally the shortest parasitic element
C. The driven element is the longest parasitic element
D. Low feed-point impedance increases bandwidth

**(B)** In a typical three-element Yagi, the director is 95% the length of the driven element and is placed at the front of the driven element in the direction signals are to be transmitted and received. The reflector element is 105% of the length of the driven element and is placed to the rear. The director is normally the shortest element of a Yagi antenna.

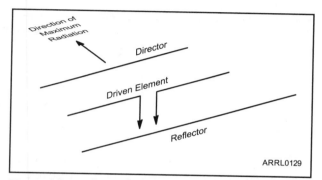

**Figure G9-5 — This drawing shows a three-element Yagi antenna.**

**G9C04** Which statement about a three-element; single-band Yagi antenna is true?

A. The reflector is normally the longest parasitic element
B. The director is normally the longest parasitic element
C. The reflector is normally the shortest parasitic element
D. All of the elements must be the same length

**(A)** In a typical three-element Yagi, the director is 95% the length of the driven element and is placed at the front of the driven element in the direction signals are to be transmitted and received. The reflector element is 105% of the length of the driven element and is placed to the rear. The director is normally the shortest element of a Yagi antenna.

**G9C05** How does increasing boom length and adding directors affect a Yagi antenna?

A. Gain increases
B. Beamwidth increases
C. Weight decreases
D. Wind load decreases

**(A)** As the boom length of a Yagi is increased and more elements are added, the directivity or gain of the antenna increases. Directivity has an advantage in that it concentrates the transmitted and received signals in the intended direction more than in other directions, thus minimizing interference and improving the signal-to-noise ratio of received signals.

**G9C06** Which of the following is a reason why a Yagi antenna is often used for radio communications on the 20 meter band?

A. It provides excellent omnidirectional coverage in the horizontal plane
B. It is smaller, less expensive and easier to erect than a dipole or vertical antenna
C. It helps reduce interference from other stations to the side or behind the antenna
D. It provides the highest possible angle of radiation for the HF bands

**(C)** A Yagi antenna provides gain or directivity. Directivity has an advantage in that it concentrates the transmitted and received signals in the intended direction more than in other directions. Directivity applies to receiving as well as transmitting, so the antenna picks up signals better from the desired direction and rejects signals coming from the sides or back of the antenna. This minimizes interference from stations in other directions. This is one important reason for using a Yagi antenna for HF operation, such as on the 20 meter band.

**G9C07** What does "front-to-back ratio" mean in reference to a Yagi antenna?

A. The number of directors versus the number of reflectors
B. The relative position of the driven element with respect to the reflectors and directors
C. The power radiated in the major radiation lobe compared to the power radiated in exactly the opposite direction
D. The ratio of forward gain to dipole gain

**(C)** Using a directional antenna helps reduce interference in that it sends the signal in the intended direction rather than off to the side or behind. Most of the radiated signal is sent in the desired direction. This is called the major lobe of the antenna's radiation pattern. A much smaller amount of the signal is radiated in other directions. If you measure the power radiated at the peak of the major lobe (or in the desired direction) and compare that with the power radiated in the exactly opposite direction, that is the antenna's "front-to-back ratio."

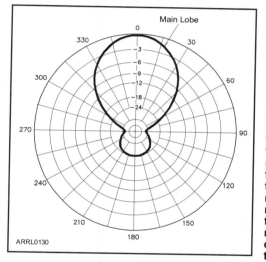

Figure G9-6 — This diagram represents the directive pattern for a typical three-element Yagi antenna. Note that this pattern is essentially unidirectional, with most of the radiation in the direction of the main lobe. There is also a small *minor lobe* at 180° from the direction of the main lobe, however. You can read the front-to-back-ratio on this type of graph by finding the strength of the minor lobe off the back of the antenna. For this antenna, the front-to-back-ratio is 24 dB because the maximum signal at 180° just touches the −24 dB circle. That means the signal off the back of the antenna is 24 dB less than the signal from the front or forward direction of the antenna.

**G9C08** What is meant by the "main lobe" of a directive antenna?

A. The magnitude of the maximum vertical angle of radiation
B. The point of maximum current in a radiating antenna element
C. The maximum voltage standing wave point on a radiating element
D. The direction of maximum radiated field strength from the antenna

**(D)** A Yagi antenna radiates most of the signal in one direction. This is called the main lobe of the antenna's radiation pattern. A much smaller amount of the signal is radiated in other directions.

**G9C09** What is the approximate maximum theoretical forward gain of a three element, single-band Yagi antenna?

A. 9.7 dBi
B. 9.7 dBd
C. 5.4 times the gain of a dipole
D. All of these choices are correct

**(A)** The maximum theoretical gain of a three-element Yagi antenna is 9.7 dBi. dBi means "decibels with respect to an isotropic antenna." An isotropic antenna is one that radiates equally in all possible directions.

**G9C10** Which of the following is a Yagi antenna design variable that could be adjusted to optimize forward gain, front-to-back ratio, or SWR bandwidth?

A. The physical length of the boom
B. The number of elements on the boom
C. The spacing of each element along the boom
D. All of these choices are correct

**(D)** All of these choices affect a Yagi antenna's forward gain, front-to-back ratio and SWR bandwidth. As you might imagine, adjusting an antenna design for the desired combination of these three important parameters can be a complicated procedure. Computer modeling programs greatly simplify this process.

**G9C11** What is the purpose of a gamma match used with Yagi antennas?

A. To match the relatively low feed-point impedance to 50 ohms
B. To match the relatively high feed-point impedance to 50 ohms
C. To increase the front to back ratio
D. To increase the main lobe gain

**(A)** A gamma match transforms the relatively low feed point impedance of a Yagi's driven element (typically 25 ohms or less) to 50 ohms to match the characteristic impedance of common coaxial feed lines.

**G9C12** Which of the following is an advantage of using a gamma match for impedance matching of a Yagi antenna to 50-ohm coax feed line?

A. It does not require that the elements be insulated from the boom
B. It does not require any inductors or capacitors
C. It is useful for matching multiband antennas
D. All of these choices are correct

**(A)** One major advantage of the gamma match is that the driven element does not have to be insulated from the antenna's boom. This simplifies mounting the element and leads to a sturdier antenna.

**G9C13** Approximately how long is each side of a quad antenna driven element?

A. ¼ wavelength
B. ½ wavelength
C. ¾ wavelength
D. 1 wavelength

**(A)** All of the elements of a quad antenna are square-shaped loops. The entire driven element of a quad antenna is approximately a full wavelength, so each of the element's four sides is approximately ¼ wavelength long.

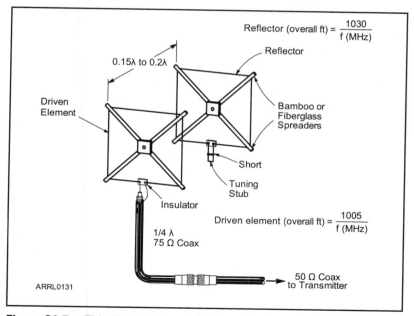

Reflector (overall ft) = $\dfrac{1030}{f\ (MHz)}$

0.15λ to 0.2λ

Reflector

Driven Element

Bamboo or Fiberglass Spreaders

Short

Tuning Stub

Insulator

Driven element (overall ft) = $\dfrac{1005}{f\ (MHz)}$

1/4 λ
75 Ω Coax

50 Ω Coax to Transmitter

ARRL0131

**Figure G9-7 — This drawing shows the construction of a quad antenna. The lengths of the driven element and reflector element are given by the equations shown in the drawing. To find the length of one side, divide these total lengths by 4.**

**G9C14** How does the forward gain of a two-element quad antenna compare to the forward gain of a three-element Yagi antenna?

A. About ⅔ as much
B. About the same
C. About 1.5 times as much
D. About twice as much

**(B)** A two-element quad (or delta loop) antenna has about the same gain as a three-element Yagi.

**G9C15** Approximately how long is each side of a quad antenna reflector element?

A. Slightly less than ¼ wavelength
B. Slightly more than ¼ wavelength
C. Slightly less than ½ wavelength
D. Slightly more than ½ wavelength

**(B)** The reflector element of a quad antenna is slightly more than a full wavelength (about 5% longer), so each of the four sides is slightly longer than ¼ wavelength. (See also Figure 9-7 and the discussion for question G9C13.)

**G9C16** How does the gain of a two-element delta-loop beam compare to the gain of a two-element quad antenna?

A. 3 dB higher
B. 3 dB lower
C. 2.54 dB higher
D. About the same

**(D)** Like the quad antenna, a delta-loop antenna uses loop elements that are approximately one wavelength long. The difference is that delta loop elements are triangle-shaped loops, rather than square. The gain of a delta loop is about the same as a quad antenna when the number of elements and spacing are the same.

**G9C17** Approximately how long is each leg of a symmetrical delta-loop antenna?

  A. ¼ wavelength
  B. ⅓ wavelength
  C. ½ wavelength
  D. ⅔ wavelength

**(B)** A delta-loop antenna has a driven element that is a triangle-shaped loop. It also uses a triangle-shaped loop reflector and sometimes one or more directors. The entire driven element of a delta-loop antenna is approximately a full wavelength, so each of the 3 sides is approximately ⅓ wavelength long.

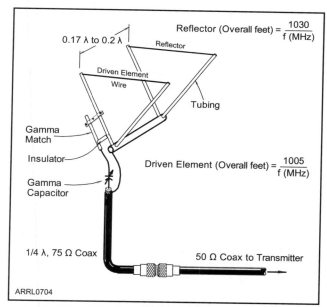

Figure G9-8 — This drawing shows the construction of a delta-loop antenna. The lengths of the driven element and reflector element are given by the equations shown in the drawing. To find the length of one side, divide these total lengths by 3.

**G9C18** What happens when the feed point of a quad antenna is changed from the center of either horizontal wire to the center of either vertical wire?

A. The polarization of the radiated signal changes from horizontal to vertical
B. The polarization of the radiated signal changes from vertical to horizontal
C. The direction of the main lobe is reversed
D. The radiated signal changes to an omnidirectional pattern

**(A)** The polarization of signals radiated by a vertically-oriented 1-wavelength loop is determined by where the feed point is located. If the feed point is located at the top or bottom of the loop, the polarization of the radiated signal is horizontal. If the feed point is located on either side of the loop, the polarization of the radiated signal is vertical. (See the discussion for G9C13.)

**G9C19** What configuration of the loops of a two-element quad antenna must be used for the antenna to operate as a beam antenna, assuming one of the elements is used as a reflector?

A. The driven element must be fed with a balun transformer
B. The driven element must be open-circuited on the side opposite the feed point
C. The reflector element must be approximately 5% shorter than the driven element
D. The reflector element must be approximately 5% longer than the driven element

**(D)** The reflector element of a quad antenna is slightly more than a full wavelength (about 5% longer), so each of the four sides is slightly longer than ¼ wavelength. (See also Figure 9-7 and the discussion for question G9C13.)

**G9C20** How does the gain of two 3-element horizontally polarized Yagi antennas spaced vertically 1/2 wavelength apart typically compare to the gain of a single 3-element Yagi?

A. Approximately 1.5 dB higher
B. Approximately 3 dB higher
C. Approximately 6 dB higher
D. Approximately 9 dB higher

**(B)** "Stacking" two Yagis the proper distance apart combines their signals so that the forward gain of the resulting array is doubled, a gain of 3 dB.

## G9D    Specialized antennas

### G9D01    What does the term "NVIS" mean as related to antennas?
A.  Nearly Vertical Inductance System
B.  Non-Visible Installation Specification
C.  Non-Varying Impedance Smoothing
D.  Near Vertical Incidence Sky wave

**(D)** NVIS, or Near Vertical Incidence Skywave, refers to a communications system that uses low, horizontally polarized antennas such as dipoles that radiate most of their signal at high vertical angles. These signals are then reflected back to Earth in a region centered on the antenna. NVIS allows stations to communicate within the skip zone for lower-angle sky wave propagation.

### G9D02    Which of the following is an advantage of an NVIS antenna?
A.  Low vertical angle radiation for working stations out to ranges of several thousand kilometers
B.  High vertical angle radiation for working stations within a radius of a few hundred kilometers
C.  High forward gain
D.  All of these choices are correct

**(B)** See question G9D01.

### G9D03    At what height above ground is an NVIS antenna typically installed?
A.  As close to one-half wave as possible
B.  As close to one wavelength as possible
C.  Height is not critical as long as it is significantly more than ½ wavelength
D.  Between ¹⁄₁₀ and ¼ wavelength

**(D)** NVIS antennas must be mounted close to the ground so that the radiated signal is maximized at high vertical angles.

### G9D04    What is the primary purpose of antenna traps?
A.  To permit multiband operation
B.  To notch spurious frequencies
C.  To provide balanced feed-point impedance
D.  To prevent out of band operation

**(A)** Traps consist of parallel LC circuits that act as electrical switches to isolate sections of the antenna at their resonant frequencies. At other frequencies, traps act as inductance or capacitance. This changes the antenna's "electrical length" automatically, allowing it to operate on two or more bands.

**G9D05** What is the advantage of vertical stacking of horizontally polarized Yagi antennas?

A. Allows quick selection of vertical or horizontal polarization
B. Allows simultaneous vertical and horizontal polarization
C. Narrows the main lobe in azimuth
D. Narrows the main lobe in elevation

**(D)** The increase in gain for a vertical stack of Yagi antennas results from narrowing the vertical width of the main lobe of a single antenna's radiation pattern. The narrower lobe results in stronger received signals and less received noise at angles away from the peak of the main lobe.

**G9D06** Which of the following is an advantage of a log periodic antenna?

A. Wide bandwidth
B. Higher gain per element than a Yagi antenna
C. Harmonic suppression
D. Polarization diversity

**(A)** A log periodic antenna is designed to provide consistent gain and feed point impedance over a wide frequency range.

**G9D07** Which of the following describes a log periodic antenna?

A. Length and spacing of the elements increases logarithmically from one end of the boom to the other
B. Impedance varies periodically as a function of frequency
C. Gain varies logarithmically as a function of frequency
D. SWR varies periodically as a function of boom length

**(A)** The name "log-periodic" refers to the ratio of length and spacing between adjacent elements of the antenna. By designing the antenna entirely in terms of ratios, the antenna's performance becomes independent of frequency over a wide range

**G9D08** Why is a Beverage antenna not used for transmitting?

A. Its impedance is too low for effective matching
B. It has high losses compared to other types of antennas
C. It has poor directivity
D. All of these choices are correct

**(B)** The Beverage antenna is used for receiving because it rejects noise and signals from unwanted directions, increasing the received signal-to-noise ratio for better copying ability. It is not used for transmitting because it is quite inefficient compared to most transmitting antennas.

**G9D09** Which of the following is an application for a Beverage antenna?

A. Directional transmitting for low HF bands
B. Directional receiving for low HF bands
C. Portable direction finding at higher HF frequencies
D. Portable direction finding at lower HF frequencies

**(B)** Beverage antennas are most effective at frequencies of 7 MHz and below. This includes the amateur MF 160 meter band, as well as the lower HF bands of 80, 60, and 40 meters.

**G9D10** Which of the following describes a Beverage antenna?

A. A vertical antenna constructed from beverage cans
B. A broad-band mobile antenna
C. A helical antenna for space reception
D. A very long and low directional receiving antenna

**(D)** The typical Beverage antenna consists of a wire 1 wavelength or more long, supported from 6 to 10 feet above ground. It is terminated with a 300 to 1000-ohm resistor connected to a ground rod at the end of the antenna in the direction of preferred reception. It is highly directional, rejecting noise and signals from unwanted directions.

**G9D11** Which of the following is a disadvantage of multiband antennas?

A. They present low impedance on all design frequencies
B. They must be used with an antenna tuner
C. They must be fed with open wire line
D. They have poor harmonic rejection

**(D)** Multiband antennas by definition are designed to radiate well on several frequencies. Most HF amateur bands are harmonically-related, meaning their frequencies are integral multiples of each other — 3.5, 7, 14, 21 and 28 MHz. While this is convenient in that one antenna can be used on separate bands, harmonics of a fundamental signal on, for example 7 MHz, will be radiated well by a multiband antenna on 14, 21 and 28 MHz.

# Electrical and RF Safety

Your General class exam (Element 3) will consist of 35 questions taken from the General class question pool as prepared by the Volunteer Examiner Coordinators' Question Pool Committee. A certain number of questions are taken from each of the 10 subelements. There will be 2 questions from the subelement shown in this chapter. These questions are divided into 2 groups, labeled G0A and G0B

## SUBELEMENT G0 — ELECTRICAL AND RF SAFETY
## [2 Exam Questions — 2 Groups]

### G0A RF safety principles, rules and guidelines; routine station evaluation

**G0A01**  What is one way that RF energy can affect human body tissue?

A. It heats body tissue
B. It causes radiation poisoning
C. It causes the blood count to reach a dangerously low level
D. It cools body tissue

**(A)** Body tissues that are subjected to very high levels of RF energy may suffer heat damage. These effects depend on the frequency of the energy, the power density of the RF field that strikes the body, and even on factors such as the polarization of the wave. The thermal effects of RF energy should not be a major concern for most radio amateurs because of the relatively low RF power we normally use and the intermittent nature of most amateur transmissions. It is rare for amateurs to be subjected to RF fields strong enough to produce thermal effects unless they are fairly close to an energized antenna or unshielded power amplifier.

**G0A02**  Which of the following properties is important in estimating whether an RF signal exceeds the maximum permissible exposure (MPE)?

A. Its duty cycle
B. Its frequency
C. Its power density
D. All of these choices are correct

**(D)** The body's natural resonant frequencies affect how the body absorbs RF energy. For this reason, polarization, power density and the frequency of the radio signal are all important in estimating the effects of RF energy on body tissue.

Electrical and RF Safety        **1**

**G0A03** How can you determine that your station complies with FCC RF exposure regulations?

A. By calculation based on FCC OET Bulletin 65
B. By calculation based on computer modeling
C. By measurement of field strength using calibrated equipment
D. All of these choices are correct

**(D)** You may use any of these three procedures to determine whether your station complies with the exposure guidelines. The simplest, by far, is to perform the calculations in FCC OET Bulletin 65. There are online and software tools to help you perform the calculations. In complex or unique situations, it may be required to model or measure the exposure. [97.13(c)(1)]

**G0A04** What does "time averaging" mean in reference to RF radiation exposure?

A. The average time of day when the exposure occurs
B. The average time it takes RF radiation to have any long-term effect on the body
C. The total time of the exposure
D. The total RF exposure averaged over a certain time

**(D)** Time averaging, when applied to RF radiation exposure, takes into account the total RF exposure averaged over either a 6-minute or a 30-minute exposure time. Time averaging compensates for the transmit/receive time ratio during normal amateur communications. It takes into account that the body cools itself after a time of reduced or no RF radiation exposure.

**G0A05** What must you do if an evaluation of your station shows RF energy radiated from your station exceeds permissible limits?

A. Take action to prevent human exposure to the excessive RF fields
B. File an Environmental Impact Statement (EIS-97) with the FCC
C. Secure written permission from your neighbors to operate above the controlled MPE limits
D. All of these choices are correct

**(A)** Some of the things you can do to prevent human exposure to excessive RF radiation are to move your antennas farther away, restrict access to the areas where exposure would exceed the limits, or reduce power to reduce the field strengths in those areas.

**G0A06** This question has been withdrawn.

**GOA07** What effect does transmitter duty cycle have when evaluating RF exposure?

A. A lower transmitter duty cycle permits greater short-term exposure levels
B. A higher transmitter duty cycle permits greater short-term exposure levels
C. Low duty cycle transmitters are exempt from RF exposure evaluation requirements
D. High duty cycle transmitters are exempt from RF exposure requirements

**(A)** Since amateurs usually spend more time listening than transmitting, low duty cycles are common. Remember that including duty cycle in the exposure evaluation takes into account the reduced average transmitted power from not operating continuously at full power. This means greater short-term exposure levels can be permitted with low-duty-cycle emissions.

**GOA08** Which of the following steps must an amateur operator take to ensure compliance with RF safety regulations when the transmitter power exceeds levels specified in Part 97.13?

A. Post a copy of FCC Part 97 in the station
B. Post a copy of OET Bulletin 65 in the station
C. Perform a routine RF exposure evaluation
D. All of these choices are correct

**(C)** Even if your station is exempt from the requirement, you may want to do a simple RF Radiation Exposure Evaluation. The results would demonstrate to yourself and possibly to your neighbors that your station is within the guidelines and is no cause for concern. None of the actions listed in the other answer choices would help to ensure that your station meets the FCC RF safety regulations.

**G0A09** What type of instrument can be used to accurately measure an RF field?

A. A receiver with an S meter
B. A calibrated field-strength meter with a calibrated antenna
C. A betascope with a dummy antenna calibrated at 50 ohms
D. An oscilloscope with a high-stability crystal marker generator

**(B)** You can use a calibrated field-strength meter and calibrated field-strength sensor (antenna) to accurately measure an RF field. Even if you have access to such an expensive laboratory-grade field-strength meter, several factors can upset the readings. Reflections from the ground and nearby conductors (power lines, other antennas, house wiring, etc.) can easily confuse field-strength readings. You must know the frequency response of the test equipment and probes, and use them only within the appropriate range. Even the orientation of the test probe with respect to the polarization of the antenna being tested is important.

**G0A10** What is one thing that can be done if evaluation shows that a neighbor might receive more than the allowable limit of RF exposure from the main lobe of a directional antenna?

A. Change from horizontal polarization to vertical polarization
B. Change from horizontal polarization to circular polarization
C. Use an antenna with a higher front-to-back ratio
D. Take precautions to ensure that the antenna cannot be pointed in their direction

**(D)** A simple way to ensure that you do not point your antenna toward a neighbor's house while you are transmitting is to clearly mark your rotator control to remind you. Some rotator controls also have programmable "no go" regions that can prevent rotating the antenna to those directions.

**G0A11** What precaution should you take if you install an indoor transmitting antenna?

A. Locate the antenna close to your operating position to minimize feed-line radiation
B. Position the antenna along the edge of a wall to reduce parasitic radiation
C. Make sure that MPE limits are not exceeded in occupied areas
D. No special precautions are necessary if SSB and CW are the only modes used

**(C)** You should locate any antenna (whether it is indoors or outdoors) as far away as practical from living spaces that will be occupied while you are operating. You should also perform a routine environmental evaluation to make sure that MPE limits are not exceeded in occupied areas.

**G0A12** What precaution should you take whenever you make adjustments or repairs to an antenna?

A. Ensure that you and the antenna structure are grounded
B. Turn off the transmitter and disconnect the feed line
C. Wear a radiation badge
D. All of these choices are correct

**(B)** One way to be sure that no one can activate the transmitter while you are working on it is to turn off the transmitter power supply and disconnect the antenna feed line. If there is a chance of anyone entering the station, it is also a good idea to post a notice that you are working on the antenna.

**G0A13** What precaution should be taken when installing a ground-mounted antenna?

A. It should not be installed higher than you can reach
B. It should not be installed in a wet area
C. It should be limited to 10 feet in height.
D. It should be installed so no one can be exposed to RF radiation in excess of maximum permissible limits

**(D)** No one should be near a transmitting antenna while it is in use. Install ground-mounted transmitting antennas well away from living areas so that people cannot come close enough to be exposed to more than the MPE limits. If there is a possibility of someone walking up to your antenna while you are transmitting, it may be a good idea to install a protective fence around the antenna.

## G0B Safety in the ham shack: electrical shock and treatment, safety grounding, fusing, interlocks, wiring, antenna and tower safety

**G0B01** Which wire or wires in a four-conductor line cord should be attached to fuses or circuit breakers in a device operated from a 240-VAC single-phase source?

A. Only the hot wires
B. Only the neutral wire
C. Only the ground wire
D. All wires

**(A)** The hot wires are the only ones that should be fused. If fuses are installed in the neutral or ground lines, an overload will open the fuses or circuit breaker but will NOT remove voltage from any equipment connected to that circuit.

**G0B02** What is the minimum wire size that may be safely used for a circuit that draws up to 20 amperes of continuous current?

A. AWG number 20
B. AWG number 16
C. AWG number 12
D. AWG number 8

(C) AWG number 12 wire is required for a 20-ampere circuit.

**G0B03** Which size of fuse or circuit breaker would be appropriate to use with a circuit that uses AWG number 14 wiring?

A. 100 amperes
B. 60 amperes
C. 30 amperes
D. 15 amperes

(D) AWG number 14 wiring should be protected by a 15-ampere fuse or circuit breaker.

**G0B04** Which of the following is a primary reason for not placing a gasoline-fueled generator inside an occupied area?

A. Danger of carbon monoxide poisoning
B. Danger of engine over torque
C. Lack of oxygen for adequate combustion
D. Lack of nitrogen for adequate combustion

(A) Carbon monoxide and other exhaust fumes can accumulate in your garage, basement or other confined living area, so ventilation is very important. Be sure not to place generators near air intakes or vents, as well.

**G0B05** Which of the following conditions will cause a Ground Fault Circuit Interrupter (GFCI) to disconnect the 120 or 240 Volt AC line power to a device?

A. Current flowing from one or more of the hot wires to the neutral wire
B. Current flowing from one or more of the hot wires directly to ground
C. Over-voltage on the hot wire
D. All of these choices are correct

(B) A GFCI opens the circuit if it detects an imbalance in the currents flowing through the hot and neutral leads. The imbalance indicates that a leakage path exists from the hot connection to safety ground. That is a serious shock hazard and should be located and repaired.

**G0B06** Why must the metal enclosure of every item of station equipment be grounded?

A. It prevents blowing of fuses in case of an internal short circuit
B. It prevents signal overload
C. It ensures that the neutral wire is grounded
D. It ensures that hazardous voltages cannot appear on the chassis

**(D)** Grounding the metal enclosure or chassis provides a path for current if there is a short circuit between the hot or neutral leads and the equipment enclosure. Without a ground connection, a shock hazard may exist if the chassis is touched.

**G0B07** Which of the following should be observed for safety when climbing on a tower using a safety belt or harness?

A. Never lean back and rely on the belt alone to support your weight
B. Always attach the belt safety hook to the belt D-ring with the hook opening away from the tower
C. Ensure that all heavy tools are securely fastened to the belt D-ring
D. Make sure that your belt is grounded at all times

**(B)** When climbing, make sure that hooks and clips are fully latched before putting any weight on them. Then be sure that for spring-loaded devices, the gate of the hook or clip cannot be opened by pressing against the tower or hardware. You might not notice that it has been accidentally unlatched!

**G0B08** What should be done by any person preparing to climb a tower that supports electrically powered devices?

A. Notify the electric company that a person will be working on the tower
B. Make sure all circuits that supply power to the tower are locked out and tagged
C. Unground the base of the tower
D. All of these choices are correct

**(B)** Before climbing, remove power from any circuit that will not be used while you are on the tower. The best way is to remove fuses or open circuit breakers. Once the circuit is opened, lock the circuit breaker open, if possible, and tag the fuse block or breaker panel so that no one will reconnect the circuit.

**G0B09** Why should soldered joints not be used with the wires that connect the base of a tower to a system of ground rods?

A. The resistance of solder is too high
B. Solder flux will prevent a low conductivity connection
C. Solder has too high a dielectric constant to provide adequate lightning protection
D. A soldered joint will likely be destroyed by the heat of a lightning strike

**(D)** When lightning strikes, the high current will melt the solder instantly and disconnect the ground wires. The proper way to bond ground wires to a tower is with a mechanical ground clamp.

Electrical and RF Safety     **7**

**G0B10**   Which of the following is a danger from lead-tin solder?
  A. Lead can contaminate food if hands are not washed carefully after handling
  B. High voltages can cause lead-tin solder to disintegrate suddenly
  C. Tin in the solder can "cold flow" causing shorts in the circuit
  D. RF energy can convert the lead into a poisonous gas

**(A)** Lead is a known toxin when ingested or inhaled. Although the amount of soldering done by most amateurs does not cause enough lead exposure to be a hazard, it is a good idea to wash your hands after soldering and not eat "at the bench."

**G0B11**   Which of the following is good engineering practice for lightning protection grounds?
  A. They must be bonded to all buried water and gas lines
  B. Bends in ground wires must be made as close as possible to a right angle
  C. Lightning grounds must be connected to all ungrounded wiring
  D. They must be bonded together with all other grounds

**(D)** Lightning protection grounds must be tied to all other safety grounds in your home and shack. Having separate ground systems can expose equipment to damage from the lightning current surge jumping between ground systems.

**G0B12**   What is the purpose of a transmitter power supply interlock?
  A. To prevent unauthorized access to a transmitter
  B. To guarantee that you cannot accidentally transmit out of band
  C. To ensure that dangerous voltages are removed if the cabinet is opened
  D. To shut off the transmitter if too much current is drawn

**(C)** High voltages are often present inside transmitter and amplifier power supplies. The interlocks on those supplies prevent you from coming in contact with energized power supply components. Interlocks often short high voltage circuits to ground when activated, providing further safety measures. Do not defeat or bypass interlock circuits unless the repair instructions specifically require you to do so.

**G0B13**   What must you do when powering your house from an emergency generator?
  A. Disconnect the incoming utility power feed
  B. Insure that the generator is not grounded
  C. Insure that all lightning grounds are disconnected
  D. All of these choices are correct

**(A)** If you do not disconnect your home's circuit breaker box from the incoming power line, called backfeeding, the power from your generator will flow back to the utility lines where it creates a shock hazard for utility workers. In addition, if utility power is restored with your generator connected to the power line, the generator may be damaged.

**GOB14** **Which of the following is covered in the National Electrical Code?**

A. Acceptable bandwidth limits
B. Acceptable modulation limits
C. Electrical safety inside the ham shack
D. RF exposure limits of the human body

**(C)** The National Electrical Code covers the wiring of electrical devices,

**GOB15** **Which of the following is true of an emergency generator installation?**

A. The generator should be located in a well ventilated area
B. The generator should be insulated from ground
C. Fuel should be stored near the generator for rapid refueling in case of an emergency
D. All of these choices are correct

**(A)** Carbon monoxide and other exhaust fumes can accumulate in your garage, basement or other confined living area, so ventilation is very important. Be sure not to place generators near air intakes or vents, as well.

**GOB16** **When might a lead-acid storage battery give off explosive hydrogen gas?**

A. When stored for long periods of time
B. When being discharged
C. When being charged
D. When not placed on a level surface

**(C)** You need to keep your lead-acid battery charged at all times for use in an emergency. A by-product of charging your battery is the release of hydrogen gas that can explode if ignited by a spark. A well-ventilated area is essential.

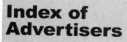

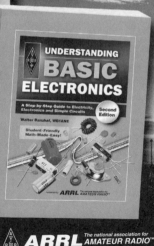

# About the ARRL ──────────────────

The seed for Amateur Radio was planted in the 1890s, when Guglielmo Marconi began his experiments in wireless telegraphy. Soon he was joined by dozens, then hundreds, of others who were enthusiastic about sending and receiving messages through the air— some with a commercial interest, but others solely out of a love for this new communications medium. The United States government began licensing Amateur Radio operators in 1912.

By 1914, there were thousands of Amateur Radio operators— hams—in the United States. Hiram Percy Maxim, a leading Hartford, Connecticut inventor and industrialist, saw the need for an organization to band together this fledgling group of radio experimenters. In May 1914 he founded the American Radio Relay League (ARRL) to meet that need.

Today ARRL, with approximately 156,000 members, is the largest organization of radio amateurs in the United States. The ARRL is a not-for-profit organization that:

- promotes interest in Amateur Radio communications and experimentation
- represents US radio amateurs in legislative matters, and
- maintains fraternalism and a high standard of conduct among Amateur Radio operators.

At ARRL headquarters in the Hartford suburb of Newington, the staff helps serve the needs of members. ARRL is also International Secretariat for the International Amateur Radio Union, which is made up of similar societies in 150 countries around the world.

ARRL publishes the monthly journal QST, as well as newsletters and many publications covering all aspects of Amateur Radio. Its headquarters station, W1AW, transmits bulletins of interest to radio amateurs and Morse code practice sessions. The ARRL also coordinates an extensive field organization, which includes volunteers who provide technical information and other support services for radio amateurs as well as communications for public-service activities. In addition, ARRL represents US amateurs with the Federal Communications Commission and other government agencies in the US and abroad.

Membership in ARRL means much more than receiving QST each month. In addition to the services already described, ARRL offers membership services on a personal level, such as the ARRL Volunteer Examiner Coordinator Program and a QSL bureau.

Full ARRL membership (available only to licensed radio amateurs) gives you a voice in how the affairs of the organization are governed. ARRL policy is set by a Board of Directors (one from each of

15 Divisions). Each year, one-third of the ARRL Board of Directors stands for election by the full members they represent. The day-to-day operation of ARRL HQ is managed by an Executive Vice President and his staff.

No matter what aspect of Amateur Radio attracts you, ARRL membership is relevant and important. There would be no Amateur Radio as we know it today were it not for the ARRL. We would be happy to welcome you as a member! (An Amateur Radio license is not required for Associate Membership.) For more information about ARRL and answers to any questions you may have about Amateur Radio, write or call:

ARRL—The national association for Amateur Radio
225 Main Street
Newington CT 06111-1494

Voice: 860-594-0200
Fax: 860-594-0259

E-mail: **hq@arrl.org**
Internet: **www.arrl.org**

Prospective new amateurs call (toll-free):
**800-32-NEW HAM** (800-326-3942)
You can also contact us via e-mail at **newham@arrl.org**
or check out the ARRL Website at **www.arrl.org**

# Notes